The Secure Information Society

T0143128

Jörg Krüger · Bertram Nickolay · Sandro Gaycken
Editors

The Secure Information Society

Ethical, Legal and Political Challenges

 Springer

Editors
Jörg Krüger
Department of Security Technology
Fraunhofer Institute for Production Systems
    and Design Technology
Berlin, Germany

Sandro Gaycken
Department of Technology, Philosophy &
    Theory
University of Stuttgart
Stuttgart, Germany

Bertram Nickolay
Department of Security Technology
Fraunhofer Institute for Production Systems
    and Design Technology
Berlin, Germany

ISBN 978-1-4471-5841-7   ISBN 978-1-4471-4763-3 (eBook)
DOI 10.1007/978-1-4471-4763-3
Springer London Heidelberg New York Dordrecht

Preface

In our modern information societies, we not only use and welcome computers. We are highly dependent upon them. This is a downside, a hubris of this kind of progress. These days, everything is done with a computer. And in most cases in which the computer replaced analogue or human routines, we forgot about those alternative practices rather quickly. This renders us even more dependent. We are not only technologically dependent. We are also cognitively dependent. Life without computers has become impossible.

If computers were a 100 % reliable, this dependency would not be a problem. But they are not reliable. They might be reliable in terms of the standard continuity of their services. This is a specification which is met these days. But they are unreliable on a different front. They are insecure. They are vulnerable to attackers. They can either be attacked directly, to disrupt their services. Or they can be abused in clever ways to do the bidding of an attacker as a dysfunctional user. This is probably more dangerous than the simple disruption. A disruption can be noticed and—in most cases—managed. An abuse, a dysfunctional use need not be noticed and can serve the attacker a host of different attractive options. Information can be stolen. It can be manipulated. Criminal goods can be transported. Vital services can be eroded. Infrastructures or products can be damaged.

Given the vast diffusion of information technologies and our high dependencies, these threats grow more dangerous every day. What is even worse—over the past years, a set of new and even more sophisticated attackers got attracted by the vulnerabilities and dependencies of modern information societies. Organized crime syndicates and militaries have realized that this kind of outset is hugely beneficial for them, with little costs to be spent for substantial monetary or strategic gains.

We have to react. But we should not react blindly. Blind reactions are likely in this field. Public criticism and the press generate a pressure to act, but decision-makers of all sorts are no experts. They cannot even decide on the right kind of expert or a truly efficient company. Success in security is notoriously hard to measure. So experts and companies with good PR-strategies tend to be in the focus of decision-makers. But those are not always the ones who are right or capable to provide the right solution. On the contrary—many IT-security products are insecure themselves, thus only adding insecurity to an already insecure system. The victim will not know for years. Neither the attacker, nor the deficient product will notify it. Concluding, the uncertainty enables security regimes with uncertain effects. Path-dependencies ensue and make it even harder to implement the security needed where it is needed. Decision-makers do not want to have made a wrong decision. So they tend to stick to whatever they have decided, and defend it for as long as possible—whether it actually provides security or not.

Implementing the wrong kind of security can even be detrimental to other political values. Freedom and security have always been in an uneasy relationship to one another. There are win-win-scenarios, and we should continue to seek them and handle them as a priority, even if they are monetarily more costly. But in many cases, freedom and security affect each other negatively. This requires even more expert knowledge. Decision-makers would not only have to be able to understand the technology, they would also have to understand its further technical and political potential, any kind of possible abuse of its functionality.

But most decision-makers still struggle with the amount of knowledge needed. And most experts struggle as well. They frequently encounter difficulties when they have to understand the implications of their field of expertise for another field of expertise. This is the burden of the complexities we live in. Technologists don't understand politics or law and vice versa.

To mitigate this problem, an approach to technological enlightenment should be initiated. More transdisciplinary knowledge has to be generated to inform lay people inasmuch as experts of other fields about the nuts and bolts of modern information societies and about the implications of technological or political progress or the lack thereof.

This is the task of this anthology. It aims to provide a spotlight onto some aspects of the uneasy relationship between information technology and information society, when it comes to security. With this general outset, it aims more narrowly to present an outlook onto some late developments and new technologies to help the dialogue not only in its current and ongoing struggle, but to anticipate the future in time and prepare perspectives for the challenges ahead. We hope you find it enlightening.

Jörg Krüger
Bertram Nickolay
Sandro Gaycken

Contents

Contributors

M. Blankenburg Technische Universität Berlin, Berlin, Germany

Chris Bronk Department of Computer Science, James A. Baker Institute for Public Policy, Rice University, Houston, USA

Johan Eriksson Department of Political Science, School of Social Sciences, Södertörn University, Huddinge, Sweden

Giampiero Giacomello Department of Political and Social Sciences, University of Bologna, Bologna, Italy

C. Horn Technische Universität Berlin, Berlin, Germany

Kristina Irion Central European University, Budapest, Hungary

Sylvia Kierkegaard International Association of IT Lawyers, Copenhagen, Denmark; University of Southampton, Southampton, UK

J. Krüger Fraunhofer Institute for Production Systems and Design Technology, Berlin, Germany

Herbert Lin National Research Council, Washington, USA

B. Nickolay Fraunhofer Institute for Production Systems and Design Technology IPK, Berlin, Germany

Hasso Plattner Enterprise Platform and Integration Concepts Chair, Hasso Plattner Institute, Potsdam, Germany

D. Pöhler Fraunhofer Institute for Production Systems and Design Technology IPK, Berlin, Germany

Matthieu-P. Schapranow Enterprise Platform and Integration Concepts Chair, Hasso Plattner Institute, Potsdam, Germany

J. Schneider Fraunhofer Institute for Production Systems and Design Technology IPK, Berlin, Germany

Nico van Eijk Institute for Information Law (IViR), University of Amsterdam, Amsterdam, The Netherlands

Part I
New Strategic Cybersecurity

Chapter 1
Between War & Peace: Considering the Statecraft of Cyberspace

Chris Bronk

Abstract This chapter considers how cyber-enabled diplomacy may be undertaken by the United States. While much discussion over the past decade has been dedicated to the topic of cyber warfare, less has attention has been directed at the use of cyber instruments (IT, social media, the blogosphere, etc.) in diplomatic engagement. Considered and critiqued here is the international cyber strategy enunciated by the Obama Administration in May 2011, regarding cyber issues and how that strategy factors into U.S. diplomatic initiatives. Covered are the: (a) emergence of cyberspace as venue for diplomacy; (b) framing of the strategy; (c) coverage of major incidents for consideration; and (d) prescriptive elements for policy development under the heading of cyber statecraft.

1.1 Introduction

On May 16, 2011, President Barack Obama invited Homeland Security Advisor John Brennan and White House Cybersecurity Coordinator Howard Schmidt to present the U.S. International Strategy for Cyberspace along with Secretary of State Hillary Clinton, Attorney General Eric Holder, Secretary of Commerce Gary Locke, Secretary of Homeland Security Janet Napolitano and Deputy Secretary of Defense William Lynn. According to Schmidt, the prescribed policy is one in which, "the United States will build an international environment that ensures global networks are open to new innovations, interoperable the world over, secure enough to support people's work, and reliable enough to earn their trust" (Schmidt 2011). It is a strategy that must cope with a world more deeply impacted by Information Technology (IT) than most policy mandarins might have accepted just a few years ago.

The research and views expressed in this paper are those of the individual researcher and do not necessarily represent the views of the James A. Baker III Institute for Public Policy or Rice University.

C. Bronk (✉)
Department of Computer Science, James A. Baker Institute for Public Policy, Rice University, Houston, USA
e-mail: rcbronk@rice.edu

J. Krüger et al. (eds.), *The Secure Information Society*, DOI 10.1007/978-1-4471-4763-3_1,   3
© Springer-Verlag London 2013

From mid-2010 moving forward to the present, IT issues and the impact of IT on international political discourse have assumed increased salience in global geopolitics (Demchak and Dombrowski 2011). The Stuxnet computer worm (Sheldon 2011a), WikiLeaks and the social-media-facilitated revolutions of the Arab Spring have already provided ample reason for a high-level U.S. policy on cyber issues. Additionally, the killing of Osama bin Laden has provided an opening for a broader strategic dialogue in Washington, one that includes cyberspace in its proper context. This policy has been a long time coming, and it has arrived in the form of the Obama administration's *International Strategy for Cyberspace* (Obama 2011) (ISC), which presents concepts and ideals on a cluster of diplomatic, commercial and security issues related to the global information space that the Internet and its environs have become.

It is important to establish what the strategy is not. It is not a cybersecurity plan, but rather a broad set of prescriptions relating to the Internet as a global construct and information more generally. (The Defense Department's July 2011, *Strategy for Operating in Cyberspace* (Panetta 2011) is more firmly directed at the security issue, however with defensive and offensive concepts.) The international cyber strategy combines defense with diplomacy and development initiatives (Pellerin 2011). Borrowing from Deibert and Rohozinski, the question we ponder is to what degree the U.S. government has decided to pursue the protection of cyberspace, and to what degree that protection holds militaristic overtones (Deibert and Rohozinski 2011). Principally, the Obama administration has enunciated a policy that places primary emphasis on the keeping the Internet a going, growing and global concern.

1.2 Maintaining a Single, Global Cyberspace

Envisaged in the Obama cyber strategy is a trustworthy Internet that remains open for business, embracing innovation and accepting entrepreneurship. This means managing cybercrime, developing international standards and keeping markets open, to the maximum degree possible across sovereign boundaries. "[A]n open, interoperable, secure and reliable cyberspace depends on nations recognizing and safeguarding that which should endure, while confronting those who would destabilize or undermine our increasingly networked world," states the document (Obama 2011, p. 3).

A trustworthy Internet is a necessity for global economic interconnection. We should see the cyber strategy as falling in line with recent contributions to the macrostrategic literature (Y 2011). There is valid concern that the United States will not be able to project its military power abroad if its strategic foundation rests upon an economic bed of sand. But the question remains as to how the Internet fits into grand strategy. Nye's conceptualization of cyber elements existing across a continuum of power options for the state is, certainly a strong starting point. So, the message communicated regarding the Internet should be clear (Nye 2011). The United States, which was a key player in creating the technologies of the Information Revolution, "is committed to preserving and enhancing the benefits of digital networks to our societies and economies" (Obama 2011, p. 4).

The values espoused by the Obama administration's cyber strategy, however, go beyond economics, spanning freedom of expression, privacy and the free flow of information. These pillars of American thinking do not scale uniformly across the planet, however. Evgeny Morozov—whose homeland, Belarus, remains under the control of Europe's perhaps last great über-authoritarian regime—notes that the Internet is no panacea for political repression (Morozov 2011). In the context of the values embraced by the strategy document, it is important to understand that the Internet by itself is not sufficient to bring about the freedoms of expression and information flow. Furthermore, the world's digitally connected population—perhaps exemplified by the 1 billion users of Facebook (Zuckerberg 2012)—have willingly abandoned a degree of privacy. While the values outlined in the ISC are clearly important, we would do well to question their uniform applicability to a global cyberspace.

President Obama has enunciated a policy that goes far beyond cybersecurity, extending outside the province of the Pentagon and its legions of information warriors.[1] In the Internet, inherited is a superb, but flawed platform for global interconnection. The problem for policy is how to develop a diplomatic dialogue with the international community on how to deliver on the strategy's lofty goals of remedying the most glaring issues of that flawed platform.[2] For example, the United States needs Russia to acquiesce to cooperation on cybercrime, and it would like the regimes of the Middle East to further open their Internet windows to the world's ideas. Perhaps most importantly, however, the U.S. will need to determine how its cyber policy will be interwoven with the greater statecraft involved in its relationship with China and that country's move toward an Internet of its own control.

How the United States handles the rise of China and its cyber insecurities (Lewis 2010), especially regarding human rights, will define what is likely the to be most important of the United States' bilateral relationships for the next several decades. Henry Kissinger argues that although the U.S. must affirm "its commitment to human dignity and popular participation in government—experience has shown that to seek to impose them by confrontation is likely to be self-defeating—especially in a country with such a historical vision of itself as China" (Kissinger 2011). This speaks to a need for international cyber policy that is nuanced, accepting of complexity and tuned to both political objectives and technical realities. One size will not fit all.

Presented by the Obama administration is a policy that will support deeply the position of U.S. economic interests, from relative upstarts Google, Facebook and Twitter as well as more established technology firms the likes of IBM, Microsoft

---

[1] On the broad concept of cyberpower, consider (Sheldon 2011b).

[2] The flaw indicated here is the Internet's development from the high-trust ARPANET research network, which was constructed to an architecture largely unconcerned with malicious actors' access to it.

and Oracle. This is important, as future prosperity rests upon the technological innovation engine that has driven the U.S. economy for decades. But at root, what will be needed is a diplomacy, indeed a *realpolitik*, that understands the value of information and the attendant dissemination mechanisms for it as well as bounds for the area in which conflict of harder nature will no doubt arise. The ICS may represent the a beginning of a geopolitical turn in which more systematic thinking is applied to international information and Internet policy, and not simply as public diplomacy, however, it is unclear how durable or applicable the strategy may be or how the forces of politics may alter the fabric of the strategy and at what pace.

1.3 Considering an International Cyber Strategy and Attendant Statecraft

As Secretary Clinton has reminded audiences in the United States and abroad, the Internet matters to America's diplomatic initiatives. This is codified in the State Department's organizational plan for the next few years, its sweeping inaugural Quadrennial Diplomacy and Development Review. One of the State Department's new capacities to be developed is in the position of a Coordinator for Cyber Issues. This position's first incumbent, Christopher Painter, has as his mandate, to "lead State's engagement on cybersecurity and other cyber issues, including efforts to protect a critical part of diplomacy—the confidentiality of communications between and among governments" (Clinton 2010). Painter's public comments remain few, although he was adamant in one public appearance, that no treaty instrument would be needed for cyberspace and that, "We need a discussion around the norms that are in cyberspace, what the rules of the road are, and we need to build a consensus around those topics" (Dodds and Satter 2011). True, consensus is needed.

Concise thinking on the "other cyber issues," however, is desirable for the U.S. foreign policy, however the capacities and deficiencies of the State Department and other agencies involved in the craft of foreign affairs, must be addressed as they move more deliberately into the international politics of the global digital information space. Considered here are the potential policy contours on cyber issues; the incident history upon which policy will be constructed; the linkages to other U.S. government entities desired for sharing in the craft of making cyber foreign policy; and the soft and hard power considerations for policy in the cyber domain.

1.4 Reading Between the Headlines: Issue Framing

No discussion on the U.S. position in the international affairs of cyberspace is possible without addressing WikiLeaks. To the author, the episode represents a leak, albeit a massive one, not a cyber attack. If we are to believe the accusations against U.S. Army intelligence specialist Bradley Manning, WikiLeaks's publication of State Department cable traffic is a case that illustrates the enormous vulnerabilities produced by the capacity of digital information to be easily copied and purloined

(The Economist 2011). A trusted insider likely caused the WikiLeaks episode, making it not entirely unlike the leaking of the Pentagon Papers.

WikiLeaks matters to this thesis as a problem of information control. Looking at regimes recently toppled in the Maghreb, the information control issue was inverted, with the problem keeping messages out rather than in. Although WikiLeaks is relevant, there have been many episodes to foretell the rising importance of cyber issues for US diplomacy,[3] from the cyber attacks launched against Estonia and Georgia (ostensibly by Russia) and those reputedly undertaken by North Korea to the fallout of the Stuxnet worm, which may have been the first precision targeted cyber attack directed against the infrastructure of Iran's nuclear fuel enrichment program (Broad 2010).

Thinking about these items in context requires more consideration of how strategy will guide policy on a multitude of issues, from encryption and distributed computing to management of Facebook and Twitter, in achievement of the international policy goals of the United States (Bendrath et al. 2007). Indeed, the entire State Department should well consider how it must re-tool virtually all of its operations to meet the digitally interconnected, transnational world in which it functions and has exhibited behavior, which demonstrates that such an activity is underway.[4] However, in Washington, cyber policy is principally a matter of cybersecurity.

1.5 Cybersecurity and U.S. Diplomacy

While control of information in Washington is as old as its politics, the importance of cyber security was discounted after September 11, 2001. Forecasts of cyber terrorism seemed misguided. Al Qaeda's attacks employed the low-tech means to achieve the dramatic of physical results. In response, rounding up jihadists became the order of the day for the Bush Administration. Nonetheless, voices from computer science and other areas continued to express concern on the cyber terror threat (Spafford 2003).

Because of the failure to connect the dots in advance of the 9/11 attacks, interconnectedness became a key issue for Washington agencies engaged in the business of international security. For the State Department, one of the top mandates to arrive before 9/11 with its then-chief, Colin Powell, was provision of Internet connectivity to all of the personal computers (Perlez 2001), both in Washington and the more than 250 diplomatic posts abroad—no small feat. Pushing information to foreign public audiences via the Internet, a task of public diplomacy, largely became the task of the Bureau of International Information Programs[5] (IIP), and its Internet

---

[3] And warnings about information security issues (Johnson 2005).

[4] This should be the raison d'etre for State's 21st Century Statecraft initiatives, some of which echo the ideas contained within Condoleezza Rice's concept of Transformational Diplomacy.

[5] Established when the United States Information Agency merged with the Department of State in 1999.

portal, USINFO.gov. The years of Powell's leadership were a period of activity in interconnection.

Whether connecting the foreign affairs workforce or digitizing public diplomacy, these are activities software developers might consider "use cases"[6] for IT, in other words applications of preexisting or developable technologies to perform work. This could be considered evolutionary application of IT to improve how international relations work was accomplished.[7] Beyond these use cases are the high politics of global cyberspace. Internet politics transcend sovereign boundaries and their stakes rise as the value added to global economic and societal well-being is attributed to IT.

Internet politics mattered little to U.S. foreign affairs even a few years ago. One of the State Department's first serious brushes with international policymaking of the Internet came in Tunis, site of the 2005 World Summit of the Information Society. There, the question was whether governance of the Internet should be transferred from the U.S.-based International Corporation on Assigned Names and Numbers (ICANN) to a transnational entity, perhaps an agency of the UN. Six years later, Tunisian protests, labeled the Jasmine Revolution, were undertaken in Tunis's streets, but also on Twitter, in blogs and on the U.S. Embassy's Facebook page. This last medium cost the State Department almost nothing to create, yet has served as a vehicle for direct communication between embassy staff and political activists, in Tunisia and beyond the country's borders.[8] The Facebook page did not belong to the State Department, *per se*, but may be widely viewed as the sovereign territory of the United States.

Events in Tunisia and later Egypt underscore the value of information technologies to the conduct of international affairs, not only between states but also publics. Micro-blogging service Twitter became a vital window into the protests following the 2009 presidential election in Iran.[9] Embassies and consulates unable to open smaller constituent posts to serve the ever growing number of global cities passing the one million inhabitant population threshold have embraced virtual presence via the Web as an alternative answer (Bain 2007). Clearly, cyberspace is a tremendous tool for both collecting information from abroad and also communicating messages from Washington to the world (Johnson 2009). It may be so much more than what the Department of Defense reduces to the term domain.

---

[6]The term "use case" is commonly utilized in IT management and software development projects. It simply is the definition of a user directed function, for example Print or Save.

[7]Considering the how IT may either provide added evolutionary improvement or revolutionize the way business is done (Varian and Shapiro 1998).

[8]The dialog is also multi-lingual, with comments in Arabic and French as well as English. Also, not to be discounted is Tunisian press coverage of leaked U.S. Embassy Tunis cables reporting excess and splendor among the ruling elite (Black 2011).

[9]When a technology update was due to be undertaken by the company, a State Department official asked for the work to be delayed so that real time information would continue to flow from Tehran and other sites of protests against the election results after internal security forces silenced foreign correspondents and removed them from the country.

1.6 Defining the Space for Coordination

Cyber issues will continue to surprise American diplomats as the foundations of political power in the Internetworked world change. Mass communication, interaction and mobilization of political expression via IT are part of something new (Shirky 2011). With regard to unfettered access to the Internet, there may be no way for national governments to stand as partially repressive. As we have seen, perhaps in most evincing example in Burma's 2007 anti-government protests, it remains possible for the most repressive of regimes to cut Internet links and black themselves out to the world.[10] How Internet connectivity impacts the internal politics of states is a rapidly evolving phenomenon. We are unable to know if the cyber cafés of the Middle East and North Africa will serve as forces for political moderation and greater democratization or as digital *madrassas* cultivating audiences prepared to employ extreme violence as the ultimate form of political expression (Wheeler 2006). This is a world in which we consider Internet freedoms issues, however, we also increasingly worry about an Internet that works and how institutions leveraging the Internet are rendered vulnerable by it.

Perhaps the clearest message sent by the ISC is that the core job at hand is the management of a secure and stable Internet. This is a task that has largely been left to the technical community. Useful suggestions can be made on securing the technical infrastructure of the Internet without engaging in a political discussion about the job at hand (Amoroso 2011). However, the question remains as to what the international political posture of the United States should be regarding the protection of information systems and conduits from unauthorized access, manipulation or disruption, and for that we simply need more fact and less speculation. Time will tell.

1.7 Push and Shove: Considering the Exemplars

For more than a decade, a number of individuals have forecast potentially devastating Internet-computer failure scenarios (Adams 2001) in which the lights go out, banking locks up, planes fall from the sky and general mayhem ensues. Some in this crowd prognosticated rhetoric of an Electronic Pearl Harbor (Munro 1995), and they remain a factor in the discourse on cyber security matters. While many have contributed to the school of deep concern regarding cyber vulnerabilities, perhaps no voice has been more important than Richard Clarke. In his *Cyber War*, Clarke (2010) caps prior work on why he considers the cyber security problem to be so important. Interesting as well is a review of the book by Bruce Schneier, a leading light in the cyber security field with strong technical credentials. Of the book, he opines,

---

[10]The Myanmar government severed its outbound Internet service within 24 hours of the killing of Japanese photojournalist Kenji Nagai on September 27, 2007 by a Burmese soldier.

> Cyber War is a fast and enjoyable read. This means you could give the book to your non-techy friends, and they'd understand most of it, enjoy all of it, and learn a lot from it. Unfortunately, while there's a lot of smart discussion and good information in the book, there's also a lot of fear-mongering and hyperbole as well. Since there's no easy way to tell someone what parts of the book to pay attention to and what parts to take with a grain of salt, I can't recommend it for that purpose. This is a pity, because parts of the book really need to be widely read and discussed (Schneier 2010).

Ultimately this speaks to the core problem for breaking down the international cyber strategy to policy; sifting through massive quantities of information to determine the real threats, the points of true international concern that require diplomatic attention and the corralling of partners and stakeholders. There is much bad behavior in cyberspace.[11] Much of it is crime, plain and simple (Wall 2007). Other activities revolve around the theft of ideas and intellectual property, from the high crimes of corporate espionage to the more pedestrian problems of peer-to-peer enabled subversion of digital rights management regimes emplaced to protect copyright, both well acknowledged forms of criminal behavior.[12] Governments too have gone from reading the mail of others to reading the email of others, and we are amply aware that the signals intelligence piece of the intelligence enterprise is enormously important (Aid 2009). There is increasing evidence of a cyber warfare *realpolitik* becoming more a reality than a hypothesis, particularly in the space of low-intensity conflict and covert action (Manjikian 2010). Finally, it appears that international law on warfare in cyberspace lags behind technical capacity (Shanker 2010). What is clear is that Internet security politics are increasingly merging with international threat politics (Dunn Cavelty 2008).

1.8 Bordering on Cyberwar: Some Hard Power Cases

While cyber operations as a component of warfare between states or in civil conflicts is a real and growing prospect, we can look to a series of events to serve as guideposts of policy on cyber misbehavior falling somewhere short of war. The cyber attacks launched against Estonia in April–May 2007[13] certainly marked an escalation in the level of impact wrought upon the function of a highly Internet-worked, digital society. While there is no definitive proof of Moscow's involvement in directing or initiating the broad denial of service attacks against Estonia,

---

[11] Charney's four categories: espionage, cybercrime, intellectual theft and cyberwar, is a useful heuristic (Charney 2010).

[12] There is a very blurry gray area regarding the line between corporate and national espionage especially considering the role of nationally-subsidized or state run companies in many sectors, including energy, aviation, and telecommunications.

[13] Cyber attacks against institutions in Estonia including telecommunications, banking and government services were precipitated by the Estonian government's decision to move the Soviet Bronze Soldier of Tallinn monument to the Great Patriotic War to a military cemetery in Tallinn's suburbs from a location in the city's core.

a NATO member, the fact remains that the attacks against Estonia were politically motivated. Whether one of Russia's cybercrime syndicates, a phenotype exemplified by the Russian Business Network (RBN), was to blame or the Russian security or intelligence services remains publicly unknown. But, Estonia's cyber incident capped more than a decade of denial of service attacks and web page defacements made between rival and belligerent states or groups, including China-Taiwan (as a corollary to cross-Straits issue), Israel-Palestine (as part of the continuing Intifada movements) and Japan-South Korea (over language regarding Korea's occupation appearing in Japanese school textbooks).

Confusion over laying blame in such cases is of course one of the key matters for international cyber security policy, the problem of attribution (Dipert 2010; Crosston 2011). The Kalashnikov and the rocket propelled grenade are the cheap, readily available tools of the insurgent, the personal computer and an Internet connection are those of the hacktivist, cyber criminal or cyber warrior. The source of an attack can be almost any machine, anywhere. Worse, major distributed attacks, which involve botnets,[14] pay little, if any, respect to sovereign geography. That means that politically motivated cyber attacks may enlist computing cycles from millions of "zombie" Internet hosts including those within the targeted country. International policy has yet to address the issue with an international agreement to ban denial of service attacks[15]—yet another issue which requires a sub-strategic policy remedy.

Also requiring attention is policy regarding the clandestine measures undertaken by the United States or its allies in cyberspace. What are we to make of a former In-Q-Tel administrative officer and counsel's statement that the Central Intelligence Agency was involved in the shipment of faulty computer controller systems to the Soviet Union that would later be involved in the largest natural gas pipeline explosion in history (Westby 2010; Safire 2004)? With this as background, the Stuxnet attack must be considered as well. Stuxnet reputedly involved the installation of malicious software code on the process control computers running the high speed centrifuges employed by the Iranian government to enrich uranium at its Natanz facility. Additional detail on what Stuxnet's functionality, drawn from its source code, is worth considering.

> In a nutshell, Stuxnet can be thought of as a stealth control system that resides on its target controllers along with legitimate program code. The ultimate goal of the attack is not the controller; it is what the controller controls. Attack code analysis reveals that the attackers had full knowledge of project, installation and instrumentation details. The attackers took great care to make sure that only their designated targets were hit. It was a marksmen's job. On target, the attack is surgical and takes advantage of deep process and equipment knowledge. The attack is not performed in a hit-and-run style, where it would be executed immediately after attaching to the controller or at the next best opportunity. Instead, the attack code carefully monitors the hijacked process for extended periods of time before executing the strike. Outputs are then controlled by Stuxnet, with neither legitimate program code nor any attached operator panel or SCADA system noticing. Stuxnet combines denial of control and denial of view, providing for the ultimate aggressive attack (Langner 2011).

---

[14] A botnet is a network of compromised computers that perform instructions clandestinely at the direction of an unauthorized party.

[15] A perfectly reasonable suggestion found in Clarke and Knake's *Cyberwar*.

This "marksmen's job," a covert action by cyber means, would seem an attractive option for a state or states, but in this case which ones? Stuxnet was an activity not without scale, complexity or considerable planning. If we are to believe the hypotheses laid out in one piece of news analysis (Broad et al. 2011), the purported Stuxnet attack on Iran's nuclear enrichment program connects Germany's Siemens,[16] who the authors conjecture collaborated with the U.S. Department of Energy on security issues in its process control computing systems, and Israel which holds the infrastructure required to integrate and test a cyber attack against an enrichment complex (the country has been enriching nuclear fuel at Dimona for several decades).

All of this is speculation, of course, however, cyber operations may serve as a preferable alternative to overt and covert uses of military force, but rendering the United States invulnerable to cyber attack, by state-sponsored groups or those without state affiliation, is an accomplishment not yet achieved (and perhaps unachievable) (Bronk 2010). The United States may possess unrivaled offensive cyber capabilities, but no doubt any would-be attacker will have plenty of targets from which to choose impacting its interests. It will be hard for the international community to take seriously American leaders who scold rivals for engaging in cyber espionage while reports emerge of major U.S. cyber attacks against threatening states or transnational groups. How covert offensive cyber force complements diplomacy will be perhaps the thorniest of issues to sort out.

Considering each of the major incidents mentioned above, U.S. diplomacy will doubtlessly cope with cyber attacks launched by international actors against one another, including allies and perhaps the United States itself. In addition, the United States, including the agencies of its federal government, will continue to be targeted by actors wishing to purloin, manipulate or deny access to information. While the worst case scenarios of cyber launched chaos will likely remain scenarios, the pattern of incidents reported over the last few years indicates significant vulnerability without remedy immediately at hand. Technical currents are merging with those of politics and policy. For U.S. international policy, this confluence of heterogeneous phenomenon will necessitate collaboration across a broad expanse of government agencies, into the private sector and reaching institutions of international governance and civil society around the globe. Managing the international security issues of the global information infrastructure will not be handled by Foggy Bottom or the Pentagon alone (Sommer and Brown 2011).

1.9 Engaging in Cyber Statecraft

Although proffering ideas of how cyber issues may be resolved may seem naïve and unsophisticated, nonetheless questions remain of when, how and where cyber polit-

---

[16]Despite numerous news stories detailing the vulnerability Stuxnet exploits in the Siemens S7-series process controllers, shares of Siemens AG rose from US$60 to over US$90 over the 52-week period ending on January 21, 2011. Stuxnet, nor any other major development in understanding SCADA vulnerability appears to have harmed the company's valuation.

ical strategies, tools and policies, a framework of cyber statecraft will take hold in Washington. For the moment, there appear two general strains of cyber power with which policy-makers will contend: hard and soft. Reflecting on the cases above, from the role of social media in the Middle Eastern revolutions of 2011 to the application of malicious software code against the Iranian nuclear program via the Stuxnet worm, there seems little point in acknowledging that both concepts are real and relevant, but to what degree? Consider Stuxnet, a malware attack on physical infrastructure that is reputed to have significantly damaged the Iranian enrichment capacity. Does such an attack qualify as an act of terrorism, war or international crime? Conversely, do the services provided by U.S. firms, such as Facebook, Twitter or Google, that may aid those wishing to overthrow foreign governments represent a serious concern to national sovereignty and security?

These are hard questions, but ones needing attention, at the Departments of Justice, State, Defense, and Homeland Security and the National Security Council as well. In the *International Strategy for Cyberspace* we have a values document. There still remains a dearth of international agreement on the running or policing of cyberspace, and not a single treaty regarding the use of cyber means for conflict in or outside of war. We are left to wonder if cyber attacks, as long as they do not kill or maim, will be illegal, but generally considered fair game, a requisite for espionage and option for covert action.

Painting such a picture appears bleak, and for good reason. The Internet has transformed human capacity for communication at a distance. Statements from U.S. officials retain a high-minded view of how it should be employed to enable transparency, combat corruption and diffuse American or Western ideals regarding speech and expression. While the causality is unclear and weighting of variables difficult, arguments that the diffusion of IT may lead foreign publics to question their political and economic surroundings appear valid. Cyber soft power, just as information-based soft power before it, whether in labeled propaganda or public diplomacy, is a real component of international relations, both bilateral and multilateral (Nye 2010). But there are unanticipated consequences of local interpretation of U.S. messages and content delivered by networks: digital, informational and social. Constructively considering concepts of some complexity, for instance the merits of representative or direct democracy, in 140-character blocks a la Twitter seems unlikely. But nonetheless the political pamphleteers of the coming decade will likely be bloggers of one sort or another.

On building policy from the foundation of strategy, the author counsels to consider constructing a more benign environment for statecraft over a polarized security space. *Much thinking has gone into how nations and others might wage cyberwar, but far less is locatable on digital diplomacy* (Dizard 2001). *Perhaps this will change if the countries, corporations and the multiplicity of others who employ, value and enjoy the Internet as a global entity are not able to do so.* As Internet boundary gateways align with sovereign boundaries, we may yet observe a fracturing of the Internet into many non-contiguous pieces. Bearing in mind Internet freedom and cybersecurity, such a fragmentation may be the most important consequence of U.S. policy poorly designed to pursue such lofty goals. There is a lot of work left on the table in translating the broad strategic announcement of the Obama Administration to a policy framework understood domestically and by international partners.

Acknowledgements This paper was aided in no small measure by conversation and correspondence with Shiu-Kai Chin, William A. Conklin, John Dinger, Edward Djerejian, Sandro Gaycken, Gary Galloway, Rex Hughes, Kamal Jabbour, Cody Monk, Stefaan Verhulst, John Villasenor and Dan Wallach.

References

Adams, J. (2001). Virtual defense. *Foreign Affairs*. May/June 2001.

Aid, M. (2009). *The secret sentry: the untold history of the national security agency*. New York: Bloomsbury Press.

Amoroso, E. (2011). *Cyber attacks: protecting national infrastructure*. Burlington: Butterworth-Heinemann.

Bain, B. (2007). Chat room diplomacy. *Federal Computer Week*. 3 September 2007.

Bendrath, R., Eriksson, J., & Giacomello, G. (2007). From 'cyberterrorist' to 'cyberwar', back and forth: how the United States securitized cyberspace. In J. Eriksson & G. Giacomello (Eds.), *International relations and security in the digital age*, Routledge: London.

Black, I. (2011). Tunisia: the WikiLeaks connection. *The Guardian*. 15 January 2011.

Broad, W. (2010). Report suggests Iran's nuclear effort has problems. *New York Times*. 24 November 2010.

Broad, W., Markoff, J., & Sanger, D. E. (2011). Israeli test on worm called crucial in Iran nuclear delay. *The New York Times*. 15 January 2011.

Bronk, C. (2010). Treasure trove or trouble: cyber-enabled intelligence and international politics. *American Intelligence Journal, 28(2)*.

Charney, S. (2010). *Rethinking the cyber threat*. Microsoft Corporation. May 2010.

Clarke, R. (2010). *Cyberwar: the next threat to national security and what to do about it*. New York: Ecco.

Clinton, H. (2010). *The first quadrennial diplomacy and development review*. Washington: U.S. Department of State, p. 7.

Crosston, M. (2011). World gone cyber MAD—how 'mutually assured debilitation' is the best hope for cyber deterrence. *Strategic Studies Quarterly*. Spring 2011.

Deibert, R., & Rohozinski, R. (2011). The new cyber military-industrial complex. *The Globe and Mail*. 28 March 2011.

Demchak, C., & Dombrowski, P. (2011). Rise of a cybered Westphalian age. *Strategic Studies Quarterly*. Spring 2011.

Dipert, R. (2010). The ethics of cyberwarfare. *Journal of Military Ethics, 9(4)*. December 2010.

Dizard, W. (2001). *Digital diplomacy: U.S. foreign policy in the information age*. Washington: Center for Strategic and International Studies.

Dodds, P., & Satter, R. (2011). *International law covers threats, cyber chief says*. Associated Press.

Dunn Cavelty, M. (2008). *Cyber-security and threat politics: US efforts to secure the information age*. London: Routledge.

Johnson, J. (2005). Cyber security at state: the stakes get higher. *Foreign Service Journal*. September 2005.

Johnson, J. (2009). The next generation. *Foreign Service Journal*. October 2009.

Kissinger, H. (2011). The China challenge. *Wall Street Journal*. 14 May 2011.

Langner, R. (2011). How to hijack a controller: why stuxnet isn't just about Siemens' PLCs. http://www.controlglobal.com/articles/2011/IndustrialControllers1101.html. Retrieved 27 January 2011.

Lewis, J. (2010). *Cyber war and competition in the China-U.S. relationship.* Remarks delivered at the China Institutes of Contemporary International Relations. 13 May 2010.

Manjikian, M. M. (2010). From global village to virtual battlespace: the colonizing of the Internet and the extension of realpolitik. *International Studies Quarterly, 54(2).*

Morozov, E. (2011). *The net delusion: the dark side of Internet freedom.* New York: Public Affairs.

Munro, N. (1995). The Pentagon's new nightmare: an electronic Pearl Harbor. *Washington Post.* 16 July 1995.

Nye, J. (2010). *Cyber power.* Belfer Center for Science and International Affairs. May 2010.

Nye, J. (2011). *The future of power.* New York: Public Affairs.

Obama, B. (2011). *International strategy for cyberspace, prosperity, security and openness in a networked world.* Washington: The White House.

Panetta, L. (2011). *Strategy for operating in cyberspace.* Washington: U.S. Department of Defense.

Pellerin, C. (2011). *White house launches international cyber strategy.* American Forces Press Service. http://www.defense.gov/news/newsarticle.aspx?id=63966.

Perlez, J. (2001). State dept.'s work rules: Powell's free and easy guide. *New York Times.* 26 January 2001.

Safire, W. (2004). The farewell dossier. *The New York Times.* 2 February 2004.

Schmidt, H. (2011). Launching the U.S. international strategy for cyberspace. *The White House Blog.* 16 May 2011.

Schneier, B. (2010). Book Review: *Cyber war, Schneier on Security.* 21 December 2010. http://www.schneier.com/blog/archives/2010/12/book_review_cyb.html. Retrieved 2 January 2011.

Shanker, T. (2010). Cyberwar nominee sees gaps in law. *New York Times.* 14 April 2010.

Sheldon, J. (2011a). Stuxnet and cyberpower in war. *World Politics Review.* 19 April 2011.

Sheldon, J. (2011b). Deciphering cyberpower, strategic purpose in peace and war. *Strategic Studies Quarterly.* Summer 2011.

Shirky, C. (2011). The political power of social media. *Foreign Affairs.* January–February 2011.

Sommer, P., & Brown, I. (2011). Reducing systemic cybersecurity risk. *Future Global Shocks, OECD.*

Spafford, E. (2003). *Cyber terrorism: the new asymmetric threat.* Testimony before the House Armed Services Committee Subcommittee on Terrorism, Unconventional Threats and Capabilities. 24 July 2003.

The Economist (2011). The leaky corporation. 26 February 2011.

Varian, H., & Shapiro, C. (1998). *Information rules: a strategic guide for the information economy.* Cambridge: Harvard University Press.

Wall, D. (2007). *Cybercrime: the transformation of crime in the information age.* Malden: Polity Press.

Westby, J. (2010). First worldwide cybersecurity summit, Dallas, Texas, 4 May 2010.

Wheeler, D. (2006). *Empowering publics: information technology and democratization in the Arab World: lessons from Internet cafes and beyond* (Oxford Internet Institute Research Report No. 11). The best answer is likely both.

Y (2011). *A national strategic narrative.* Woodrow Wilson International Center for Scholars.

Zuckerberg, M. (2012). One billion people on Facebook. 4 October 2012. http://newsroom.fb.com/News/One-Billion-People-on-Facebook-1c9.aspx. Accessed 19 October 2012.

Chapter 2
Laying an Intellectual Foundation for Cyberdeterrence: Some Initial Steps

Herbert Lin

Abstract This paper considers the basic question of how to effectively prevent, discourage, and inhibit hostile activity against important U.S. information systems and networks. It contains four main sections (Sections 2.1–2.3 of this paper are essentially a reproduction of The NRC letter report for the committee on deterring cyberattacks: informing strategies and developing options for U.S. policy, available at http://www.nap.edu/openbook.php?record_id=12886&page=2, 2010. Section 2.4 is based on material contained in National Research Council, in Proceedings of a workshop on deterring cyberattacks: informing strategies and developing options for U.S. policy, 2010). Section 2.1 describes a broad context for cybersecurity, establishing its importance and characterizing the threat. Section 2.2 sketches a range of possible approaches for how the nation might respond to cybersecurity threats, emphasizing how little is known about how such approaches might be effective in an operational role. Section 2.3 describes a research agenda intended to develop more knowledge and insight into these various approaches. Section 2.4 provides a summary of 15 papers by individual authors that address various aspects of the research agenda.

2.1 The Broad Context for Cybersecurity[1]

An important policy goal of the United States is to prevent, discourage, and inhibit hostile activity against the important information technology systems of the United States. This paper considers the threat of cyberattack, which refer to the deliberate use of cyber operations—perhaps over an extended period of time—to alter, disrupt,

---

[1] The discussion in this section is based on Chap. 1 of National Research Council (2009) and Chap. 2 of National Research Council (2007).

H. Lin (✉)
National Research Council, Washington, USA
e-mail: hlin@nas.edu

J. Krüger et al. (eds.), *The Secure Information Society*, DOI 10.1007/978-1-4471-4763-3_2,   17
© Springer-Verlag London 2013

deceive, degrade, usurp, or destroy adversary computer systems or networks or the information and/or programs resident in or transiting these systems or networks.[2] Cyberattack is not the same as cyber exploitation, which is an intelligence-gathering activity rather than a destructive activity and refers to the use of cyber operations—perhaps over an extended period of time—to support the goals and missions of the party conducting the exploitation, usually for the purpose of obtaining information resident on or transiting through an adversary's computer systems or networks.

Cyberattack and cyber exploitation are technically very similar, in that both require a vulnerability, access to that vulnerability, and a payload to be executed. They are technically different only in the nature of the payload to be executed. These technical similarities often mean that a targeted party may not be able to distinguish easily between a cyber exploitation and a cyberattack.

Because of the ambiguity of cyberattack and cyber exploitation from the standpoint of the targeted party, the term "cyberintrusion" will be used to refer to a hostile cyber activity where the nature of the activity is not known (that is, an activity that could be either a cyberattack or a cyber exploitation).

The range of possibilities for cyberintrusion is quite broad. A cyberattack might result in the destruction of relatively unimportant data or the loss of availability of a secondary computer system for a short period of time—or it might alter top-secret military plans or degrade the operation of a system critical to the nation, such as an air traffic control system, a power grid, or a military command and control system. Cyber exploitations might target the personal information of individual consumers or critical trade secrets of a business, military war plans, or design specifications for new weapons. Although all such intrusions are worrisome, some of these are of greater significance to the national well-being than others.

Intrusions are conducted by a range of parties, including disgruntled or curious individuals intent on vandalizing computer systems, criminals (sometimes criminal organizations) intent on stealing money, terrorist groups intent on sowing fear or seeking attention to their causes, and nation-states for a variety of national purposes. Nation-states can tolerate, sponsor, or support terrorist groups, criminals, or even individuals as they conduct their intrusions. A state might tolerate individual hackers who wish to vandalize an adversary's computer systems, perhaps for the purpose of sowing chaos. Or it might sponsor or hire criminal organizations with special cyber expertise to carry out missions that it did not have the expertise or the capability to undertake. Or it might provide support to terrorist groups by looking the other way as those groups use the infrastructure of the state to conduct Internet-based operations. In times of crisis or conflict, a state might harbor (or encourage, or control, or fail to discourage) "patriotic hackers" or "cyber patriots" who conduct hostile cyberintrusions against a putative adversary. Note that many such actions would also be plausibly deniable by the government of the host state.

---

[2]This report does not consider the use of electromagnetic pulse (EMP) attacks. For a comprehensive description of the threat from EMP attacks, see *Report of the Commission to Assess the Threat to the United States from Electromagnetic Pulse (EMP) Attack*, available at http://www.globalsecurity.org/wmd/library/congress/2004_r/04-07-22emp.pdf.

The threats that adversaries pose can be characterized along two dimensions—the sophistication of the intrusion and the damage it causes. Though these two are often related, they are not the same. Sophistication is needed to penetrate good cyberdefenses, and the damage an intrusion can cause depends on what the adversary does after it has penetrated those defenses. As a general rule, a greater availability of resources to the adversary (e.g., more money, time, talent) will tend to increase the sophistication of the intrusion that can be launched against any given target and thus the likelihood that the adversary will be able to penetrate the target's defenses.

Two important consequences follow from this discussion. First, because nation-state adversaries can bring to bear enormous resources to conduct an intrusion, the nation-state threat (perhaps conducted through intermediaries) is the most difficult to defend against. Second, stronger defenses reduce the likelihood but cannot eliminate the possibility that even less sophisticated adversaries can cause significant damage.

2.2 A Range of Possibilities

The discussion below focuses primarily on cyberattacks as the primary policy concern of the United States, and addresses cyber exploitation as necessary.

2.2.1 The Limitations of Passive Defense and Some Additional Options

The central policy question is how to achieve a reduction in the frequency, intensity, and severity of cyberattacks on U.S. computer systems and networks currently being experienced and how to prevent the far more serious attacks that are in principle possible. To promote and enhance the cybersecurity of important U.S. computer systems and networks (and the information contained in or passing through these systems and networks), much attention has been devoted to passive defense—measures taken unilaterally to increase the resistance of an information technology system or network to attack. These measures include hardening systems against attack, facilitating recovery in the event of a successful attack, making security more usable and ubiquitous, and educating users to behave properly in a threat environment (National Research Council 2007).

Passive defenses for cybersecurity are deployed to increase the difficulty of conducting the attack and reduce the likelihood that a successful attack will have significant negative consequences. But experience and recent history have shown that they do not by themselves provide an adequate degree of cybersecurity for important information systems and networks.

A number of factors explain the limitations of passive defense. As noted in previous NRC reports (National Research Council 2002, 2007), today's decision-making calculus regarding cybersecurity excessively focuses vendor and end-user attention on the short-term costs of improving their individual cybersecurity postures to the detriment of the national cybersecurity posture as a whole. As a result, much of the critical infrastructure on which the nation depends is inadequately protected against cyberintrusion.

A second important factor is that passive defensive measures must succeed every time an adversary conducts a hostile action, whereas the adversary's action need succeed only once. Put differently, attacks can be infinitely varied, whereas defenses are only as strong as their weakest link. This fact places a heavy and asymmetric burden on a defensive posture that employs only passive defense.

Because passive defenses do not eliminate the possibility that an attack might succeed, it is natural for policy makers to seek other mechanisms to deal with threats that passive defenses fail to address adequately. Policy makers understandably aspire to a goal of preventing cyberattacks (and cyber exploitations as well), but most importantly to a goal of preventing *serious* cyberattacks—cyberattacks that have a disabling or a crippling effect on critical societal functions on a national scale (e.g., military mission readiness, air traffic control, financial services, provision of electric power). In this context, "deterrence" refers to a tool or a method used to help achieve this goal. The term "deterrence" itself has a variety of connotations, but broadly speaking, deterrence is a tool for dissuading an adversary from taking hostile actions.

Adversaries that might conduct cyberintrusions against the United States span a broad range and may well have different objectives. Possible adversaries include nation-states that would use cyberattacks to collect intelligence, steal technology, or "prepare the battlefield" for use of cyberattacks either by themselves or as part of a broader effort (perhaps involving the use or threat of use of conventional force) to coerce the United States; sophisticated elements within a state that might not be under the full control of the central government (e.g., Iranian Revolutionary Guards); criminal organizations seeking illicit monies; terrorist groups operating without state knowledge; and so on.

In principle, policy makers have a number of approaches at their disposal to further the broad goal of preventing serious cyberattacks on the United States. In contrast to passive defense, all of these approaches depend on the ability to attribute hostile actions to specific responsible parties (although the precise definition of "responsible party" depends to a certain extent on context).

The first approach, and one of the most common, is the use of law enforcement authorities to investigate cyberattacks, and then identify and prosecute the human perpetrators who carry out these attacks. Traditionally, law enforcement actions serve two purposes. First, when successful, they remove such perpetrators from conducting further hostile action, at least for a period of time. Second, the punishment imposed on perpetrators is intended to dissuade other possible perpetrators from conducting similar actions. However, neither of these purposes can be served if the cyberattacks in question cannot be attributed to specific perpetrators.

In a cyber context, law enforcement investigations and prosecutions have had some success, but the time scale on which such activities yield results is typically on the order of months, during which time cyberattacks often continue to plague the victim. As a result, most victims have no way to stop an attack that is causing ongoing damage or loss of information. In addition, the likelihood that any given attack will be successfully investigated and prosecuted is low, thus reducing any potential deterrent effect. Notwithstanding the potential importance of law enforcement activities for the efficacy of possible deterrence strategies, law enforcement activities are beyond the scope of this report and will not be addressed further herein.

A second approach relies on deterrence as it is classically understood. The classical model of deterrence (discussed further in Sect. 2.2.2) seeks to prevent hostile actions through the threat of retaliation or responsive action that imposes unacceptable costs on a potential adversary or denies an adversary the benefits that may result from taking those hostile actions. Deterrence thus includes active defense, in which actions can be taken to neutralize an incoming cyberattack.

A third approach takes note of the fact that the material threat of retaliation underlying deterrence is not the only method of inhibiting undesirable behavior. Behavioral restraint (discussed further in Sect. 2.2.3) is more often the result of formal law and informal social norms, and the burden of enforcement depends a great deal on the robustness of such rules and the pressures to conform to those rules that can be brought to bear through the social environment that the various actors inhabit.

These approaches—and indeed an approach based on passive defense—are by no means mutually exclusive. For example, some combination of strengthened passive defenses, deterrence, law enforcement, and negotiated behavioral restraint may be able to reduce the likelihood that highly destructive cyberattacks would be attempted and to minimize the consequences if cyberattacks do occur. But how well any of these approaches can or will work to prevent cyberattacks (or cyberintrusions more broadly) is open to question, and indeed is a topic in need of serious research.

2.2.2 Classical Deterrence[3]

Many analysts have been drawn to the notion of deterring hostile activity against important IT systems and networks, rather than just defending against such activity. Deterrence seems like an inevitable choice in an offense-dominant world—that is, a world in which offensive technologies and tactics are generally capable of thwarting defensive efforts. As noted in Sect. 2.2.1, a major difficulty of defending against hostile actions in cyberspace arises from the asymmetry of offense versus defense.

Deterrence was and is a central construct in contemplating the use of nuclear weapons and in nuclear strategy. Because effective defenses against nuclear weapons are difficult to construct, using the threat of retaliation to persuade an adversary to refrain from using nuclear weapons is regarded by many as the most plausible and effective alternative to ineffective or useless defenses. Indeed, deterrence

---

[3]The discussion in Sect. 2.2.2 is based on Chap. 9 of National Research Council (2009).

of nuclear threats in the Cold War establishes the paradigm in which the conditions for successful deterrence are largely met.

Although the threat of retaliation is not the only possible mechanism for practicing deterrence, such a threat is in practice the principal and most problematic method implied by use of the term.[4] Extending traditional deterrence principles to cyberattack (that is, cyberdeterrence) would suggest an approach that seeks to persuade adversaries to refrain from launching cyberattacks against U.S. interests, recognizing that cyberdeterrence would be only one of a suite of elements of U.S. national security policy.

But it is an entirely open question whether cyberdeterrence is a viable strategy. Although nuclear weapons and cyber weapons share one key characteristic (the superiority of offense over defense), they differ in many other key characteristics, and the section below discusses cyberdeterrence and when appropriate contrasts cyberdeterrence to Cold War nuclear deterrence. What the discussion below will suggest is that nuclear deterrence and cyberdeterrence do raise many of the same questions, but indeed that the answers to these questions are quite different in the cyber context than in the nuclear context.

The U.S. Strategic Command formulates deterrence as follows (U.S. Department of Defense 2006):

> Deterrence [seeks to] *convince adversaries* not to take *actions that threaten U.S. vital interests* by means of decisive influence over their decision-making. Decisive influence is achieved by *credibly threatening* to *deny benefits* and/or *impose costs*, while *encouraging restraint* by convincing the actor that restraint will result in an *acceptable outcome*.

For purposes of this report, the above formulation will be used to organize the remainder of this section, by discussing at greater length the words in italics above. Nevertheless, there are other plausible formulations of the concept of deterrence, and these formulations might differ in tone and nuance from that provided above.

2.2.2.1 Convince

At its root, convincing an adversary is a psychological process. Classical deterrence theory assumes that actors make rational assessments of costs and benefits and refrain from taking actions where costs outweigh benefits. But it assumes unitary actors (i.e., a unitary decision maker whose cost-benefit calculus is determinative for all of the forces under his control), and also that the costs and benefits of each actor are clear, well-defined, and indeed known to all other actors involved, and further that these costs and benefits are sufficiently stable over time to formulate and implement a deterrence strategy. Classical deterrence theory bears many similarities

---

[4] Analysts also invoke the concept of deterrence by denial, which is based on the prospect of deterring an adversary through the prospect of failure to achieve its goals—facing failure, the adversary chooses to refrain from acting. But denial is—by definition—difficult to practice in an offense-dominant world.

to neoclassical economics, especially in its assumptions about the availability of near-perfect information (perfect in the economic sense) about all actors.

Perhaps more importantly, real decisions often take place during periods of crisis, in the midst of uncertainty, doubt, and fear that often lead to unduly pessimistic assessments. Even a cyberattack conducted in peacetime is more likely to be carried out under circumstances of high uncertainty about the effectiveness of technology on both sides, the motivations of an adversary, and the effects of an attack.

In addition, cyber conflict is relatively new, and there is not much known about how cyber conflict would or could evolve in any given situation. History shows that when human beings with little hard information are placed into unfamiliar situations in a general environment of tension, they often substitute supposition for knowledge. In the words of a former senior administration official responsible for protecting U.S. critical infrastructure, "I have seen too many situations where government officials claimed a high degree of confidence as to the source, intent, and scope of a [cyber]attack, and it turned out they were wrong on every aspect of it. That is, they were often wrong, but never in doubt" (National Research Council 2009, p. 142).

As an example, cyber operations that would be regarded as unfriendly during normal times may be regarded as overtly hostile during periods of crisis or heightened tension. Cyber operations X, Y, and Z undertaken by party A (with a history of neutrality) may be regarded entirely differently if undertaken by party B (with a history of acting against U.S. interests). Put differently, reputations and past behavior matter—how we regard or attribute certain actions that happen today will depend on what has happened in the past.

This point has particular relevance as U.S. interest in obtaining offensive capabilities in cyberspace becomes more apparent. The United States is widely regarded as the world leader in information technology, and such leadership can easily be seen by the outside world as enabling the United States to conceal the origin of any offensive cyber operation that it might have conducted. That is, many nations will find it plausible that the United States is involved in any such operation against it, and even if no U.S.-specific "fingerprints" can be found, such a fact can easily be attributed to putative U.S. technological superiority in conducting such operations.

Lastly, a potential adversary will not be convinced to refrain from hostile action if it is not aware of measures the United States may take to retaliate. Thus, some minimum of information about deterrence policy must be known and openly declared. This point is further addressed in Sect. 2.2.2.4.

2.2.2.2 Adversaries

In the Cold War paradigm of nuclear deterrence, the world is state-centric and bipolar. It was reasonable to presume that only nation-states could afford to assemble the substantial infrastructure needed to produce the required fissile material and develop nuclear weapons and their delivery vehicles. That infrastructure was sufficiently visible that an intelligence effort directed at potential adversaries could keep track of the nuclear threat that possible adversaries posed to the United States. Today's concerns about terrorist use of nuclear weapons arise less from a fear that terrorists will develop and build their own nuclear weapons and more from a fear that they will be able to obtain nuclear weapons from a state that already has them.

These characteristics do not apply to the development of weapons for cyberattack. Many kinds of cyberattack can be launched with infrastructure, technology, and background knowledge easily and widely available to nonstate parties and small nations. Although national capabilities may be required for certain kinds of cyberattack (such as those that involve extensive hardware modification or highly detailed intelligence regarding truly closed and isolated system and networks), substantial damage can be inflicted by cyberattacks based on ubiquitous technology.

A similar analysis holds for identifying the actor responsible for an attack. In the nuclear case, an attack on the United States would have been presumed to be Soviet in origin because the world was bipolar. In addition, surveillance of potential launch areas provided high-confidence information regarding the fact of a launch, and also its geographical origin—a missile launch from the land mass of any given nation could be safely attributed to a decision by that nation's government to order that launch.

Sea-based or submarine-based launches are potentially problematic in this regard, although in a bipolar world, the Soviet Union would have been deemed responsible. In a world with three potential nuclear adversaries (the United States, Soviet Union, and China), intensive intelligence efforts have been able to maintain to a considerable extent the capability for attributing a nuclear attack to a national power, through measures such as tracking adversary ballistic missile submarines at sea. Identification of the distinctive radiological signatures of potential adversaries' nuclear weapons is also believed to have taken place.

The nuclear deterrence paradigm also presumes unitary actors, nominally governments of nation-states—that is, it presumes that the nuclear forces of a nation are under the control of the relevant government, and that they would be used only in accordance with the decisions of national leaders.

These considerations do not hold for cyberattack, and for many kinds of cyberattack the United States would almost certainly not be able to ascertain the source of such an attack, even if it were a national act, let alone hold a specific nation responsible. For example, the United States is constantly under cyberattack today, and it is widely believed (though without conclusive proof) that most of these cyberattacks are not the result of national decisions by an adversary state, though press reports have claimed that some are.

In general, prompt technical attribution of an attack or exploitation—that is, identification of the responsible party (individual? subnational group? nation-state?) based only on technical indicators associated with the event in question—is quite problematic, and any party accused of launching a given cyberintrusion could deny it with considerable plausibility. Forensic investigation might yield the identity of the responsible party, but the time scale for such investigation is often on the order of weeks or months. (Although it is often quite straightforward to trace an intrusion to the proximate node, in general, this will not be the origination point of the intrusion. Tracing an intrusion to its actual origination point past intermediate nodes is what is most difficult.)

Three factors mitigate to some (unknowable) degree this bleak picture regarding attribution. First, for reasons of its own, a cyberattacker may choose to reveal to its target its responsibility for a cyberattack. For example, it may conduct a cyberattack of limited scope to demonstrate its capability for doing so, acknowledge its responsibility, and then threaten to conduct a much larger one if certain demands are not met.[5]

Second, over time a series of cyberintrusions might be observed to share important technical features that constitute a "signature" of sorts. Thus, the target of a cyberattack may be able to say that it was victimized by a cyberattack of type X on 16 successive occasions over the last 3 months. An inference that the same party was responsible for that series of attack might under some circumstances have some plausibility.

Third, the target of a cyberattack may have nontechnical information that points to a perpetrator, such as information from a well-placed spy in an adversary's command structure or high-quality signals intelligence. If such a party reports that the adversary's forces have just launched a cyberattack against the United States, or if a generally reliable communications intercept points to such responsibility, such information might be used to make a plausible inference about the state responsible for that attack. Political leaders in particular will not rely only on technical indicators to determine the state responsible for an attack—rather, they will use all sources of information available to make the best possible determination.

Nevertheless, it is fair to say that absent unusually good intelligence information, high confidence in the attribution of a cyberattack to a nation-state is almost certain to be unattainable during and immediately after that attack, and may not be achievable for a long time afterward. Thus, any retaliatory response to a cyberattack using either cyber or kinetic weaponry may carry a significant risk of being directed improperly, perhaps with grave unintended consequences.

2.2.2.3 Actions that Threaten U.S. Vital Interests

What actions is the United States trying to deter, and would the United States know that an action has occurred that threatens its vital interests?

A nuclear explosion on U.S. territory is an unambiguously large and significant event, and there is little difficulty in identifying the fact of such an explosion. The United States maintains a global network of satellites that are capable of detecting

---

[5]Of course, a forensic investigation might *still* be necessary to rule out the possibility that the putative attacker was only claiming responsibility for the attack when in fact it had no real ability to conduct the attack on its own. To mitigate the possibility that it might not be believed, the party claiming responsibility could leave a "calling card" in the wake of an attack whose contents only it could know.

and locating nuclear explosions in the air and on the ground, and a network of seismic sensors that provide additional information to localize nuclear explosions. Most importantly, a nuclear explosion would occur against the very quiet background of zero nuclear explosions happening over time.

But U.S. computer and communications systems and networks are under constant cyberintrusion from many different parties, and against this background noise, the United States would have to notice that critical systems and networks were being attacked and damaged. A cyberattack on the United States launched by an adversary might target multiple sites—but correlating information on attacks at different sites against a very noisy background to determine a common cause is today technically challenging. Target sets may be amorphous and complex, especially when massively complex and globally scaled supply chains are involved. And the nature of a questionable event (an intrusion) is often in doubt—is it an attack or an exploitation? If an attack, does a destructive cyberattack take place when the responsible software agent is *implanted* in a critical U.S. system, or when it is *activated*? Even knowing the effect or impact of an attack or exploitation is difficult, as the consequences of some intrusions will play out only over an extended period of time. (For example, an attack may be designed to have no immediate impact and only later to show destructive consequences.)

Another profound difference between the nuclear and cyber domains is that nuclear weapons are not thought to target individual private sector entities—it would be highly unusual for a major corporation, for example, to be the specific target of a nuclear weapon. By contrast, major corporations are subject to cyberattacks and cyber exploitations on a daily basis. This difference raises the question of whether deterrence of such intrusions on individual private sector entities (especially those that are regarded as a part of U.S. critical infrastructure) is an appropriate goal of U.S. policy—as suggested by recent allegations of Chinese cyberintrusions against human rights activists using Google's gmail.com service and against multiple private sector companies in the United States seeking important intellectual property of these companies (Cha and Nakashima 2010). The question is important, because targeted private entities might seek to defend themselves by retaliating against attackers or cyber spies, notwithstanding criminal prohibitions, with consequences damaging to U.S. national interests.

The question is important for a number of reasons. First, U.S. military forces have not been used in recent years to support the interests of specific private sector entities, at least not as a matter of declared public policy. Thus, an explicit threat to respond with force, whether cyber or otherwise, to a cyberattack on an individual private sector entity would constitute a major change in U.S. policy. Second, targeted private entities might seek to defend themselves by retaliating against attackers or cyber spies, even though such actions are currently illegal under U.S. law, and such retaliation by these entities might well have consequences damaging to U.S. national interests.

2.2.2.4 Credible Threat

A credible threat is one that an adversary believes can and will be executed with a sufficiently high probability to dissuade the adversary from taking action. (The definition of "sufficiently high" is subject to much debate and almost certainly depends on the specific case or issue in question. In some cases, even a low absolute probability of executing the deterrent threat is sufficient to dissuade.) In the nuclear domain, the United States developed strategic forces with the avowed goal of making them survivable regardless of what an adversary might do. Survivability means that these forces will be able to execute the retaliatory threat for which they are responsible under any possible set of circumstances. In addition, the United States conducts many highly visible military training exercises involving both its conventional and nuclear forces, at least in part to demonstrate its capabilities to potential adversaries.

On the other hand, U.S. capabilities for offensive cyber operations are highly classified, at least in part because discussing these capabilities in the open may point the way for adversaries to counter them. That is, at least some capabilities for conducting offensive cyber operations depend on a vulnerability that an adversary would be able to fix, if only he knew about it. To the extent that U.S. capabilities for cyber operations are intended to be part of its overall deterrent posture, how should the United States demonstrate those capabilities? Or is such demonstration even necessary given widespread belief in U.S. capabilities?

A credible deterrent threat need not be limited to a response in kind—the United States has a wide variety of options for responding to any given cyberattack, depending on its scope and character; these options include a mix of changes in defense postures, law enforcement actions, diplomacy, economic actions, cyberattacks, and kinetic attacks.[6]

Another dimension of making a threat credible is to communicate the threat to potential adversaries. A nation's declaratory policy underpins such communication and addresses, in very general terms, why a nation acquires certain kinds of weapons and how those weapons might be used. For example, a declaratory policy of the United States regarding nuclear weapons was stated in the National Military Strategy of 2004 (Joint Chiefs of Staff 2004):

> Nuclear capabilities [of the United States] continue to play an important role in deterrence by providing military options to deter a range of threats, including the use of WMD/E and large-scale conventional forces. Additionally, the extension of a credible nuclear deterrent to allies has been an important nonproliferation tool that has removed incentives for allies to develop and deploy nuclear forces.

---

[6]Chapter 1 of National Research Council (2009). As illustrations, a change in defensive posture might include dropping low-priority services, installing security patches known to cause inconvenient but manageable operational problems, restricting access more tightly, and so on. Law enforcement actions might call for investigation and prosecution of perpetrators. Diplomacy might call for demarches delivered to a perpetrator's government or severing diplomatic relations. Economic actions might involve sanctions.

For the use of cyber weapons, the United States has no declaratory policy, although the DOD Information Operations Roadmap of 2003 stated that "the USG should have a declaratory policy on the use of cyberspace for offensive cyber operations."[7]

Lastly, a "credible threat" may be based on the phenomenon of blowback, which refers to a bad consequence affecting the instigator of a particular action. In the cyberattack context, blowback may entail direct damage caused to one's own computers and networks as the result of a cyberattack that one has launched. For example, if Nation X launched a cyberattack against an adversary using a rapidly multiplying but uncustomized and indiscriminately targeted worm over the Internet, the worm might return to adversely affect Nation X's computers and networks. Blowback might also refer to indirect damage—a large-scale cyberattack by Nation X against one of its major trading partners (call it Nation Y) that affected Nation Y's economic infrastructure might have effects that could harm Nation X's economy as well. If concerns over such effects are sufficiently great, Nation X may be deterred (more precisely, self-deterred) from conducting such attacks against Nation Y (or any other major trading partner). Blowback may sometimes refer to counterproductive political consequences of an attack—for example, a cyberattack launched by a given government or political group may generate a populist backlash against that government or group if attribution of the attack can be made to the party responsible.

For blowback to be the basis of a credible threat, the dependencies that give rise to blowback should be apparent (or at least plausible) to a potential attacker. (As a possible example, it may be that given massive Chinese investment in U.S. securities, the Chinese have a large stake in the stability of U.S. financial markets, and thus might choose to refrain from an attack that might do significant harm to those markets.)

2.2.2.5 Denying Benefits

The ability to deny an adversary the benefits of an attack has two salutary results. First, an attack, if it occurs, will be futile and not confer on the adversary any particular advantage. Second, if the adversary believes (in advance) that he will not gain the hoped-for benefits, he will be much less likely to conduct the attack in the first place.

In the nuclear domain, ballistic missile defenses are believed to increase the uncertainty of an attack's success. For this reason, they need not be perfect—only good enough to significantly complicate an adversary's planning to the point at which it becomes impossible to carry out an attack with a high probability of success.

In the cyber domain, a number of approaches can be used to deny an adversary the benefits of an attack. Passive defenses can be strengthened in a number of ways, such as reducing the number of vulnerabilities present in vital systems, reducing

---

[7] Available at http://www.gwu.edu/~nsarchiv/NSAEBB/NSAEBB177/info_ops_roadmap.pdf.

the number of ways to access these systems, configuring these systems to minimize their exposed security vulnerabilities, dropping traffic selectively, and so on. Properties such as rapid recoverability or reconstitution from a successful attack can be emphasized.

Active defense may also be an option. Active defense against an incoming cyberattack calls for an operation, usually a cyber operation, that can be used to neutralize that incoming attack. A responsive operation (often described within the U.S. military as a "computer network defense response action") must be conducted while the adversary's cyberattack is in progress, so that there is an access path back to the facilities being used to mount the attack. In practice, active defense is possible only for certain kinds of cyberattack (e.g., denial-of-service attacks) and even then only when the necessary intelligence information on the appropriate targets to hit is available to support a responsive operation.

On the other hand, whether improvements in denying benefits are sufficient to deter a cyber adversary is open to question. Experience to date suggests that strengthening a system's passive defense posture may discourage the casual attacker, but will only suffice to delay a determined one. That is, the only costs to the attacker result from the loss of time and thus an increased uncertainty about its ability to conduct a successful attack on a precise timetable. Such uncertainty arguably contributes to deterrence if (and only if) the action being deterred is a necessary prelude to some other kind of attack that must also be planned and executed along a particular timetable.

2.2.2.6 Imposing Costs

Costs that may be imposed on an adversary typically involve the loss of assets or functionality valued by the adversary.

In the nuclear case, the ability to attribute an attack to a national actor, coupled with a knowledge of which specific states are nuclear-capable, enables the United States to identify target sets within each potential nuclear adversary, the destruction of which the United States believes would be particularly costly to those adversaries.

In the context of cyberattack, an attacker determined to avoid U.S. retaliation may well leave a false trail for U.S. forensic investigators to follow; such a trail would either peter out inconclusively or even worse, point to another nation that might well see any U.S. action taken against it as an act of war. (Catalytic conflict, in which a third party instigates mutual hostilities between two nations, is probably much easier in cyberspace than in any other domain of potential conflict.)

That said, the ability to attribute political responsibility for a given cyberattack is the central threshold question.

If responsibility cannot be attributed, the only hope of imposing any costs at all lies in identifying an access path to the platforms involved in launching the cyberattack on U.S. interests. For example, if it is possible to identify an access path to the attacking platforms in the midst of an ongoing cyberattack, knowledge of the national (or subnational) actor's identity may not be necessary from a technical perspective to neutralize those platforms. (An analogy would be an unidentified airplane dropping bombs on a U.S. base—such an airplane could be shot down without knowing anything about the airplane or its pilot other than the fact that it was dropping bombs on a U.S. base.) Under these circumstances, a strike-back has some chance of neutralizing an incoming cyberattack even if the identity of the adversary is not known. By developing capabilities to deny the adversary a successful cyberattack through neutralization, the United States might be able to deter adversaries from launching at least certain kinds of cyberattack against the United States. Yet neutralization is likely to be difficult—destroying or degrading the source of a cyberattack while the attack is in progress may simply lead the adversary to launch the attack from a different source. It is also extremely likely that the attacking platforms will belong to innocent parties.

The attacking platforms may also be quite inexpensive—personal computers can be acquired for a few hundred dollars, and any software used to conduct an attack is virtually free to reproduce. Thus, the attacking platforms may not be assets that are particularly valuable to the attacker. Intermediate nodes that participate in an attack, such as the subverted computers of innocent parties used in a botnet, cost nothing from a capital standpoint, although they do represent some non-zero cost to the attacker of electronically capturing and subverting them.

The location(s) of the attacking platforms may be valuable to the attacker—more precisely, keeping such locations secret may be important to the attacker. But an adversary that chooses to conduct a cyberattack using platforms located in a particular location has also probably made the choice that he is willing to lose that secret location.

If responsibility can be attributed to a known actor, the range of possibilities for response becomes much larger. For example, if a nation-state can be identified as being responsible, anything of value to that state can be attacked, using any available means.[8] Indeed, options for responding to cyberattacks span a broad range and include a mix of changes in defensive postures, law enforcement actions, diplomacy, economic actions, and kinetic attacks, as well as cyberattacks.[9] Further, if

---

[8]One particular option deserves mention along these lines. As noted earlier, the U.S. Joint Chiefs of Staff wrote in 2004 that "Nuclear capabilities... [provide] military options to deter a range of threats, including the use of WMD/E and large-scale conventional forces." The same document defines WMD/E as follows: "The term WMD/E relates to a broad range of adversary capabilities that pose potentially devastating impacts. WMD/E includes chemical, biological, radiological, nuclear, and enhanced high explosive weapons as well as other, more asymmetrical 'weapons.' They may rely more on disruptive impact than destructive kinetic effects. For example, cyberattacks on U.S. commercial information systems or attacks against transportation networks may have a greater economic or psychological effect than a relatively small release of a lethal agent." Although the use of nuclear weapons against a known adversary could indeed impose very substantial costs, the threat to use nuclear weapons in response to any kind of cyberattack on the United States would not be credible to all adversaries.

[9]Some of these potential responses are less escalatory (e.g., changes in defensive postures); others, more so (e.g., retaliatory cyberattacks or kinetic attacks). Implementing less escalatory responses

individual/personal responsibility can be ascertained (or narrowed to a sufficiently small group of individuals), severe penalties could also be imposed, ranging from law enforcement prosecutions to permissible kinetic responses.

A variety of considerations might apply to choosing the appropriate retaliatory mode. For example, a "tit-for-tat" retaliatory response against an adversary might call for a cyberattack of comparable scale against a comparable target. However, a threat to do so might not be credible if the United States has a great deal to lose from such an action, thus throwing doubt on the viability of an "in-kind" deterrence strategy. On the other hand, a near-peer competitor might well be deterred from launching a large-scale cyberattack by the knowledge that it too would have much to lose if the United States launched an in-kind counterattack.

It may even be the case that when the responsible party is known, a responsive cyberattack is among the least useful tools for responding. Because a cyber adversary knows the time of his cyberattack, he can take action to mitigate the costs that the United States will attempt to impose following his attack. For example, the adversary can take steps in advance to invalidate the intelligence information on cyber targets that the defender has already collected on him, thus strengthening its defensive posture. Such an action could force the United States into either a nonselective retaliation or a retaliation delayed until new intelligence information can be collected. In the first case, the United States may not be willing to risk the large-scale escalation that might accompany a non-selective retaliatory cyberattack, and in the second case, the adversary may have already achieved its objectives by the time a new retaliatory strike can be planned.

Whether the *prompt* imposition of costs is necessary for deterrence is another unknown. U.S. nuclear forces and their command and control are structured to support prompt responses (in part because of a "use-it-or-lose-it" concern not necessarily present in a cyber context), and such a structure is believed to be an important element of deterring nuclear attack against the United States.

By contrast, the relationship between the pace at which responses are made and the deterrent effect of such responses in a cyber context is not well understood. Although a prompt response to an incoming cyberattack may have a number of possible benefits (e.g., a demonstration of resolve, an earlier termination of the damage resulting from an attack), such a response also raises the risk that a response may be misdirected or even undertaken mistakenly. There may be more to gain by seeking more information and being more confident about the necessary attributions.

2.2.2.7 Encouraging Restraint

Under the Cold War paradigm of nuclear deterrence, the technical prerequisite to encourage restraint on an adversary's part was the ability to execute a devastating

---

would seem to require lower levels of authority than would more escalatory responses, and thus would be more easily undertaken.

response no matter what the adversary did first. In particular, the existence of a powerful ballistic missile submarine force was regarded as the element of force structure that precluded a successful counterforce first strike by an adversary. More abstractly, it was the existence of a secure second-strike capability that was the foundation of encouraging restraint on the adversary's part.

In the cyber environment, there appears to be no realistic possibility of a targeted counterforce attack that will eliminate a nation's ability to execute offensive operations in cyberspace. Cyberattack forces are too easily dispersed (indeed, can operate covertly in other nations) and can launch attacks from myriad venues. (A broad and indiscriminate attack on the Internet infrastructure—analogous to a countervalue strike—might make it hard to mount a response in kind, at least until Internet services were restored.)

But it is still an open question if a secure second-strike cyberattack capability is an enabling condition for encouraging restraint on an adversary's part. That is, does the existence of a secure U.S. cyberattack capability contribute materially to encouraging an adversary to refrain from conducting offensive operations against the United States in cyberspace? Or could other U.S. capabilities for responding compensate for any shortfall in U.S. cyberattack capabilities? A related question is whether U.S. cyberattack capabilities contribute to deterring hostile adversary actions outside cyberspace. In this context, pre-emption to eliminate an adversary's cyberattack capabilities does not seem likely or plausible, although U.S. cyberattack capabilities could be used to disrupt an adversary's impending kinetic attack.

Restraint is also a concept that is relevant to escalation after conflict has begun. That is, after conflict has broken out (whether in cyberspace or kinetically), policy makers will seek to deter an adversary from escalating the conflict to greater levels of violence. In general, deterring escalation requires that the adversary believe that escalation will result in a worse outcome than maintaining the status quo, which implicitly requires that the United States have reserve capabilities (whether cyber or kinetic) that can produce such an outcome.

2.2.2.8 Acceptable Outcome

Whatever else it may be, an acceptable outcome surely involves a cessation of hostilities. A cessation of hostilities necessarily involves the transmission of orders from the cognizant political authority to its "shooters" to refrain from undertaking further offensive actions. A reciprocal or mutual cessation of hostilities involves both sides taking such action, and one party's cessation is generally conditional on the other side's cessation. Each party must therefore be convinced that the other side has ceased or will cease hostilities.

When conventional or nuclear conflict is involved, a cessation of hostilities is reasonably easy to recognize—no more missiles fly, no more nuclear weapons explode, and so on. But when cyber conflict is involved, recognizing a cessation of hostilities is quite problematic.

For example, given that there exists a background level of ongoing cyberattacks affecting the United States, how would the United States recognize that an adversary had ceased its cyberattacks? What evidence would be acceptable as proof positive that an adversary was complying with a cyber cease-fire?

Cessation of hostilities may also call for the removal of destructive elements emplaced in an adversary's information technology infrastructure. For example, if the United States had implanted Trojan horse software agents useful for cyberattack in an adversary's infrastructure, it might be obliged to remove them or render them harmless under the terms of a cease-fire. This could entail either some direct communications between the United States and these agents (which could be monitored and thus could reveal sensitive operational secrets of the United States) or keeping track of where such agents were implanted. Autonomous attack agents that require no further command direction after deployment and replicate themselves as they spread through adversary networks are particularly problematic in this regard.

Finally, both sides may have actors under their nominal jurisdiction that do not necessarily respond to national decisions to cease and desist. For example, in the aftermath of the August 2001 incident in which a Chinese fighter airplane was destroyed and a U.S. reconnaissance airplane forced to land on Chinese territory, private individuals on each side (so-called "patriotic hackers") began to conduct cyberattacks against various web sites of the other. In ordinary kinetic hostilities, private individuals do not generally have the physical wherewithal to participate directly in combat operations. But where cyberattack is concerned, they often do, and "combat operations" takes on an expanded meaning of "operations that damage or destroy adversary information technology or information."

2.2.2.9 Observations About Cyberdeterrence

An analysis of cyberdeterrence as traditionally conceived requires a knowledge of the specific adversary being deterred, the undesirable action to be deterred, the specific threat that constitutes the basis for deterrence, and the target(s) against which the threat is to be exercised.[10] These factors are not independent—for example, the nature of the relevant specific threat and target set for effective deterrence of a nation-state may well be different than that for a terrorist group, because what is both valuable and vulnerable to the former adversary (e.g., targets of economic significance) may not be to the latter (which does not have targets of economic significance and may not care if such targets are destroyed in its host nation). In short, a generalized cyberdeterrence strategy that does not account for individual adversaries and hostile actions is less likely to succeed than one that is appropriately tailored. Of course, the price for tailored deterrence is high—a great deal of knowledge and intelligence about specific adversaries is necessary to execute such a strategy.

---

[10] See Box 9.1 (National Research Council 2009).

Where cyberattacks launched by nation-states are at issue, cyberdeterrence should not be conceptualized as being necessarily separate from other spheres of potential conflict. Although it is possible that conflict between nations might occur entirely within cyberspace, there is no reason to presume that a sufficiently serious cyberattack would not have consequences in physical space. One reason, of course, is that computer systems and the physical world often do interact—computer systems control physical artifacts and accept data from the physical world. Adversary cyberattacks may also be accompanied by other hostile behavior, such as kinetic attacks or adverse economic actions.

The threats that are at the center of deterrence need not be limited to in-kind responses. Options for responding to cyberattacks on the United States span a broad range and include a mix of changes in defensive postures, law enforcement actions, diplomacy, cyberattacks, and kinetic attacks, and there is no reason that a retaliatory cyberattack would necessarily be favored over a retaliatory kinetic attack.

There is also a broad range of conflict scenarios to which cyberdeterrence may be applicable. For example, analysts often refer to strategic or tactical conflict between adversaries. A large-scale use of cyberattack against the critical infrastructure of a nation (e.g., against its electric grid, against its financial systems) might well be regarded as strategic in nature, whereas a cyberattack against an air defense radar system would almost certainly be regarded as tactical. Such different scenarios, or scenarios located at any point along this continuum of potentially deterrable cyberattacks, may well pose different challenges for how and to what extent deterrence is relevant to them. (For example, there may well be differences in the nature of the relevant deterrent threat or the likelihood that the deterrent threat would be carried out.)

The feasibility of cyberdeterrence and of international regimes to constrain cyberattacks on the United States is profoundly affected by the fact that the technology for cyberattacks is broadly and inexpensively available to everyone, nation-states and subnational entities down to the level of single individuals. Such broad availability means that the assumption of unitary actors is not necessarily valid.

Furthermore and as mentioned in Sect. 2.2.2.4, an environment in which certain critical infrastructures are highly interconnected across national boundaries leaves open a possibility (of unknown magnitude) that a cyberattack conducted in one nation may have global effects, including effects on the instigating nation. Perhaps the most prominent example is the existence of myriad cross-border links between financial institutions, and the consequent possibility that the U.S. financial sector (for example) might be harmed from an attack against another country's financial system.

Lastly, the private sector has a direct stake in U.S. cyberattack policy—uniquely more so than for policy regarding most other kinds of military action because of the extent of private sector ownership and operation of many of the national critical infrastructure systems that must be protected. In addition, to the extent that policy needs require certain cyberattacks to be carried out, private sector cooperation may well be required. (At the very least, accidental or inadvertent interference with a U.S. government cyberattack will have to be avoided.) And as noted in Sect. 2.2.2.3, questions arise about whether deterrence of cyberattacks against individual private sector entities is properly a component of U.S. policy. An answer in the affirmative will raise the question of whether granting private sector entities the right to engage in active defense as a response to cyberattacks directed at them would enhance or detract from cyberdeterrence.

2.2.3 International Regimes that Limit or Require Certain Behaviors

The preceding discussion suggests that at the very least, classical deterrence theory (as construed for deterring nuclear attacks on the United States) is quite problematic when applied to cyberattacks on the United States because many of the conditions necessary for nuclear deterrence are absent from the cyber domain.

Whether a deterrence framework can be developed for the cyber domain is open to question, and indeed is one primary subject of the papers to be commissioned for this project. But whatever the useful scope for deterrence, there may also be a complementary and helpful role for international legal regimes and codes of behavior designed to reduce the likelihood of highly destructive cyberattacks and to minimize the realized consequences if cyberattacks do occur. That is, participation in international agreements may be an important aspect of U.S. policy.

In the past, nations have pursued a variety of agreements intended to reduce the likelihood of conflict and to minimize the realized consequences if conflict does occur (and also to reduce the financial costs associated with arms competitions) under the broad rubric of arms control. To achieve these objectives, arms control regimes often seek to limit capabilities of the signatories or to constrain the use of such capabilities. Thus, in the nuclear domain, agreements have (for example) been reached to limit the number and type of nuclear weapons and nuclear weapons platforms of the signatories—a limitation on capability that putatively reduces the destructiveness of conflict by limiting the capabilities on each side.

Agreements have also been reached for purposes of constraining the use of such capabilities—for example, the United States and Russia are parties to an agreement to provide advance notice to each other of a ballistic missile launch. Other proposed restrictions on use have been more controversial—for example, nations have sometimes sought agreement on "no first use of nuclear weapons." Agreements constraining the use of such capabilities are intended to reduce the possibility of misunderstandings that might lead to conflict and thus reduce the likelihood of conflict.

Lastly, international legal regimes and codes of behavior can make certain kinds of weapons unacceptable from a normative standpoint. For example, most nations today would eschew the overt use of biological weapons, and thus the likelihood of such use by any of these nations is lower than it would be in the absence of such a behavioral norm.

In the present case (that is, in thinking about ways to prevent cyberattacks of various kinds), one of the most powerful rationales for considering international agreements in the cyber domain is that all aspects of U.S. society, both civilian and military, are increasingly dependent on information technology, and to the extent that such dependencies are greater for the United States than for other nations, restrictions on cyberattack asymmetrically benefit the United States. Proponents of such agreements also argue that aggressive pursuit of cyberattack capabilities will legitimize cyberattack as a military weapon and encourage other nations to develop such capabilities for use against the United States and its interests, much to its detriment.

Objections to such regimes usually focus on the difficulty (near-impossibility) of verifying and enforcing such an agreement. But the United States is a party to a number of difficult-to-enforce and hard-to-verify regimes that regulate conflict and prescribe rules of behavior—notably the Biological Weapons Convention (BWC). In recent years, the BWC has been criticized for lacking adequate verification provisions, and yet few policy makers suggest that the convention does not further U.S. interests.

In the cyber domain, meaningful agreements to limit acquisition of cyberattack capability are unlikely to be possible. Perhaps the most important impediment to such agreements is the verification issue—technology development for cyberattack and the testing of such technology would have few signatures that could be observed, even with the most intrusive inspection regimes imaginable.

Agreements to constrain cyberattack capabilities are also problematic, in the sense that little can be done to verify that a party to such an agreement will in fact restrict its use when it decides it needs to conduct a cyberattack. On the other hand, such agreements have a number of benefits.

- They help to create international norms regarding the acceptability of such behavior (and major nation-states tend to avoid engaging in broadly stigmatized behavior).
- They help to inhibit training that calls for such use (though secrecy will shield clandestine training).
- The violation of such agreements may be detectable. Specifically, cyberattacks that produce small-scale effects may be difficult to detect, but massively destructive attacks would be evident from their consequences, especially with appropriate rules to assist forensic assessment. If a violation is detected, the violator is subject to the consequences that follow from such detection.

Lastly, even though the development of regimes constraining use would address only cyberattacks associated with nation-states, they could have significant benefit, as nation-states do have advantages in pursuing cyberattack that most nonstate-supported actors do not have. Although such regimes would not obviate the need for passive defenses, they could be useful in tamping down risks of escalation and might help to reduce international tensions in some circumstances.

As illustrations of regimes constraining use, nations might agree to confidence-building measures that committed them to providing mutual transparency regarding their activities in cyberspace, to cooperate on matters related to securing cyberspace (e.g., in investigating the source of an attack), to notify each other regarding certain activities that might be viewed as hostile or escalatory, or to communicate directly with each other during times of tension or crisis. Agreements to eschew certain kinds of cyberattack under certain circumstances could have value in reducing the likelihood of kinetic conflict in those cases in which such cyberattacks are a necessary prelude to a kinetic attack.

Limitations on cyber targeting (e.g., no cyberattacks on civilian targets; requirements that military computers be explicitly identified; no first use of cyberattack on a large scale; or no attacks on certain classes of targets, such as national power grids, financial markets or institutions, or air traffic control systems) could prevent or reduce the destructiveness of an attack, assuming that collateral and/or cascading damage could be limited. Agreements (or unilateral declarations) to abide by such agreements might be helpful in establishing appropriate rules of conduct (norms of behavior) and a social structure to enforce those rules.

On the other hand, U.S. policy makers and analysts have not seriously explored the utility and feasibility of international regimes that deny the legitimacy of cyberattacks on critical infrastructure assets, such as power grids, financial markets, and air traffic control systems.[11] How useful would such a regime be, especially applied in concert with a significantly improved cyberdefensive posture for these assets? How would difficulties of verification and enforcement affect relative national military postures and the credibility of the regime? What meaningful capabilities would the United States be giving up if it were to agree to such a regime? These and other related questions find few answers in the literature. The feasibility of these or other regimes to limit use of cyberattack is unclear, especially in light of the difficulties of working out the details of how the regime would actually operate. It is for this reason that research is needed to explore their feasibility.

Agreements in a cyber context might also usefully address important collateral issues, such as criminal sanctions or compensation for damages sustained under various circumstances. They might also require signatories to pass national laws that criminalize certain kinds of cyber behavior undertaken by individuals and to cooperate with other nations in prosecuting such behavior, much as the Convention on Cyber Crime has done.[12]

There are a number of major complications associated with arms control regimes for cyberattack. These include:

- The functional similarity between cyber exploitation and cyberattack. That is, from the target's perspective, it may be difficult or impossible to distinguish between a cyber operation intended for attack and one intended for exploitation. Restrictions on cyberattack will almost certainly restrict cyber exploitation to a large degree, and nations—including the United States—may well be loath to surrender even in principle any such capability for gaining intelligence.

---

[11] Indeed, the United States has until recently avoided discussions on military uses of cyberspace. In December 2009, it was publicly reported that the United States had begun to engage with Russian officials and with UN officials (see Markoff and Kramer 2009), although the emphasis of the United States in these talks was apparently directed toward combating Internet crime and as a collateral effect strengthening defenses against any militarily-oriented cyberattacks.

[12] See http://conventions.coe.int/Treaty/EN/Treaties/html/185.htm.

- The lack of state monopoly over cyber weapons. For kinetic weaponry, the destructiveness and potency of any given weapon has some significant correlation with the extent to which it is only available to nation-states—almost everyone has access to rifles, whereas jet fighters and submarines are mostly restricted to nations. For cyber weapons, this correlation is far less strong, and private parties can and do wield some cyber weapons that can be as destructive and powerful as some of those wielded by nation-states. Although as a rule nation-states do have major operational advantages in conducting cyberattacks (e.g., intelligence agencies that can support cyberattack), nonstate actors are certainly capable of acquiring cyber weaponry that can cause enormous damage.
- "Positive inspection" arrangements to increase the confidence that each side is abiding by an agreement not to engage in proscribed activities could be easily thwarted or circumvented. One primary reason is that the footprint of personnel and equipment needed to conduct cyber operations is small, and thus could be located virtually anywhere in a nation (or even in another nation).
- In contrast to nuclear weapons, the private sector has essentially unlimited access to most of the technology that underlies cyberattack weapons, and the scope for destructive use varies over a much wider range. Thus, an extraordinary degree of intrusiveness would be required to impose controls on the private acquisition and use of cyber weapons. It would be impractical and unacceptable, not to mention futile, to subject every personal computer and all forms of electronic communication to inspection to ensure that cyber weapons are not present on computers or concealed within e-mails. On the other hand, special rules might help to regulate access to the operations of critical social infrastructure in order to improve the attribution of parties that come into contact with them.
- The inherent anonymity of cyberattacks, mentioned above, greatly complicates the attribution of responsibility for an attack, and thus it is difficult to hold violators of any agreement accountable. Any alleged violation could simply be met with a strongly worded denial, and unambiguous evidence supporting the allegation would be hard to provide. Moreover, behavioral norms are generally much harder to instill and enforce in an environment in which actors can act anonymously.

Suggestions are often made to create a parallel Internet (call it an SAI, for strongly authenticated Internet) that would provide much stronger authentication of users than is required on today's Internet and would in other ways provide a much more secure environment.[13] If important facilities, such as power grids and financial institutions, migrated to an SAI, accountability for misbehavior would be much greater (because of the lack of anonymity) and the greater security of the

---

[13]For example, the White House Cyberspace Policy Review of May 2009 called for the nation to "implement, for high-value activities (e.g., the Smart Grid), an opt-in array of interoperable identity management systems to build trust for online transactions" (White House 2009). More recently, a trade press article reported on the intent of the Defense Information Systems Agency of the U.S. Department of Defense to establish an enclave for its unclassified networks that is isolated from public Internet access (Corrin 2010).

environment would mean that only very sophisticated parties could mount attacks on it or within it.

Although the availability of an SAI would certainly improve the security environment over that of today, it is not a panacea. Perhaps most importantly, SAI users would immediately become high-priority targets to be compromised by nontechnical cyberattacks. A compromised SAI user would then become an ideal platform from which to launch IT-based cyberattacks within the SAI—and in particular, would become an ideal jumping-off point for slowly and quietly assembling an array of computing resources that can be used for attack—all of which would be on the SAI. In addition, experience with large networks indicates that maintaining an actual air-gap isolation between an SAI and the standard Internet or dial-up or wireless connections would be all but impossible—not for technical reasons but because of a human tendency to make such connections for the sake of convenience.

- Subnational groups can take action independently of governments. Subnational groups may be particularly difficult to identify, and are likely to have few if any assets that can be targeted. Some groups (such as organized hacker groups) regard counterattacks as a challenge to be welcomed rather than a threat to be feared. Finally, a subnational group composed of terrorists or insurgents might seek to provoke retaliation in order to galvanize public support for it or to provoke anti-American sentiments in its supporting public.

This last point is particularly relevant to any international agreements or regime that the United States might deem helpful in reducing cyberattacks against it—any legal agreement or regime must be respected by all parties, including the United States. If the United States wishes other nations to eschew certain actions or to abide by certain behavioral requirements or to grant it certain rights under certain circumstances, it too must be willing to do the same with respect to other nations.

As an example, some analysts have suggested that it is an appropriate strategy for the United States to seek the right to retaliate against a nation for offensive acts emanating from within its borders, even if that nation's government denies responsibility for those attacks and asserts that those responsible are nonstate actors. Doing so, they argue, would give states an incentive to crack down on harmful private offensive actors in its borders. On the other hand, it is not clear that it is in the U.S. interest for the United States to be subject to such a regime, given that parties within the United States are themselves responsible for conducting many cyberattacks against the rest of the world. Any solution proposed for other nations must (most probably) be tolerable to the United States as well, but accepting such consequences may be politically, or economically, or legally infeasible.

It should also be noted that the traditional arms control agreements are not the only form of agreement that might be helpful (National Research Council 2009, Chap. 10). For example, nations have sometimes agreed on the need to protect some area of international activity such as airline transport, telecommunications, maritime activities, and so on, and have also agreed on standards for such protection.

They may declare certain purposes collectively with regard to a given area of activity on which they agree, often in the form of a multilateral treaty, and then establish consensus-based multilateral institutions (generally referred to as "specialized agencies" composed of experts rather than politicians) to which to delegate (subject to continuous review) the task of implementing those agreed purposes.

It has sometimes been easier to obtain agreement among the nations involved on standards and methods concerning the civilian (commercial) aspects of a given activity than to obtain agreement on the military (governmental) aspects of the same activity (Sofaer and Goodman 2000). For example, civil aviation is regulated internationally through agencies that have promulgated numerous agreements and regulations, all by consensus. Over the years, some precedents, and some forms of regulation, have been established, again largely by consensus, that have enhanced the protection of civilian aviation and reduced the uncertainties regarding governmental (military) aviation. A similar pattern of international regulation has resulted in increased maritime safety.

In both areas, states have agreed to criminalize terrorist attacks, and to prosecute or extradite violators. These commitments have not uniformly been kept, but security has been enhanced in these areas of international commerce because of the virtually universal support given to protecting these activities from identified threats. It is an open question whether such an approach might enhance cybersecurity internationally, whether or not it excludes any direct application or restriction on the national security activities of signatories.

2.2.4 Domestic Regimes to Promote Cybersecurity

Law enforcement regimes to prosecute cyber criminals are not the only ones possible to help promote cybersecurity. As noted in *Toward a Safer and More Secure Cyberspace*, the nation's cybersecurity posture would be significantly enhanced if all owners and operators of computer systems and networks took actions that are already known to improve cybersecurity. That is, the nation needs to do things that the nation already knows how to do.

What that report identified as a critical problem in cybersecurity was a failure of action. That report attributed the lack of adequate action to two factors—the fact that decision makers discount future possibilities of disaster so much that they do not see the need for present-day action (that is, they weigh the immediate costs of putting into place adequate cybersecurity measures, both technical and procedural, against the potential future benefits (actually, avoided costs) of preventing cyber disaster in the future—and systematically discount the latter as uncertain and vague) and the additional fact that the costs of inaction are not borne by the relevant decision makers (that is, the nation as a whole bears the cost of inaction, whereas the cost of action is borne by the owners and operators of critical infrastructure, which are largely private-sector companies).

Accordingly, that report called for changes in the decision-making calculus that at present excessively focuses vendor and end-user attention on the short-term costs of improving their cybersecurity postures. The report did not specify the nature of the necessary changes, but rather noted the need for more research in this area to assess the pros and cons of any given change.

The present report reiterates the importance of changing the decision-making calculus described above, but suggests that developing the necessary domestic regime (including possibly law, regulation, education, culture, and norms) to support a new calculus will demand considerable research.

2.3 A Possible Research Agenda

Although the preceding section seeks to describe some of the essential elements of cyberdeterrence, it is sobering to realize the enormity of intellectually unexplored territory associated with such a basic concept. Thus, considerable work needs to be done to explore the relevance and applicability of deterrence and prevention/inhibition to cyber conflict. At the highest level of abstraction, the central issue of interest is to identify what combinations of posture, policies, and agreements might help to prevent various actors (including state actors, nonstate actors, and organized criminals) from conducting cyberattacks that have a disabling or a crippling effect on critical societal functions on a national scale (e.g., military mission readiness, air traffic control, financial services, provision of electric power).

The broad themes described below (lettered A-H) are intended to constitute a broad forward-looking research agenda on cyberdeterrence. Within each theme are a number of elaborating questions that are illustrative of those that would benefit from greater exploration and analysis. Thoughtful research and analysis in these areas would contribute significantly to understanding the nature of cyberdeterrence.

A. Theoretical Models for Cyberdeterrence

1. Is there a model that might appropriately describe the strategies of state actors acting in an adversarial manner in cyberspace? Is there an equilibrium state that does not result in cyber conflict?
2. How will any such deterrence strategy be affected by mercenary cyber armies for hire and/or patriotic hackers?
3. How does massive reciprocal uncertainty about the offensive cyberattack capabilities of the different actors affect the prospect of effective deterrence?
4. How might adversaries react technologically and doctrinally to actual and anticipated U.S. policy decisions intended to strengthen cyberdeterrence?
5. What are the strengths and limitations of applying traditional deterrence theory to cyber conflict?
6. What lessons and strategic concepts from nuclear deterrence are applicable and relevant to cyberdeterrence?
7. How could mechanisms such as mutual dependencies (e.g., attacks that cause actual harm to the attacker as well as to the attacked) and counterproductivity (e.g., attacks that have negative political consequences against the attacker) be

used to strengthen deterrence? How might a comprehensive deterrence strategy balance the use of these mechanisms with the use of traditional mechanisms such as retaliation and passive defense?

B. Cyberdeterrence and Declaratory Policy

8. What should be the content of a declaratory policy regarding cyberintrusions (that is, cyberattacks and cyberintrusions) conducted against the United States? Regarding cyberintrusions conducted by the United States? What are the advantages and disadvantages of having an explicit declaratory policy? What purposes would a declaratory policy serve?
9. What longer-term ramifications accompany the status quo of strategic ambiguity and lack of declaratory policy?
10. What is the appropriate balance between publicizing U.S. efforts to develop cyber capabilities in order to discourage/deter attackers and keeping them secret in order to make it harder for others to foil them?
11. What is the minimum amount and type of knowledge that must be made publicly available regarding U.S. government cyberattack capabilities for any deterrence policy to be effective?
12. To the extent that a declaratory policy states what the United States will not do, what offensive operational capabilities should the United States be willing to give up in order to secure international cooperation? How and to what extent, if at all, does the answer vary by potential target (e.g., large nation-state, small nation-state, subnational group, and so on)?
13. What declaratory policy might help manage perceptions and effectively deter cyberattack?

C. Operational Considerations in Cyberdeterrence

14. On what basis can a government determine whether a given unfriendly cyber action is an attack or an exploitation? What is the significance of mistaking an attack for an exploitation or vice versa?
15. How can uncertainty and limited information about an attacker's identity (i.e., attribution), and about the scope and nature of the attack, be managed to permit policy makers to act appropriately in the event of a national crisis? How can overconfidence or excessive needs for certainty be avoided during a cyber crisis?
16. How and to what extent, if at all, should clear declaratory thresholds be established to delineate the seriousness of a cyberattack? What are the advantages and disadvantages of such clear thresholds?
17. What are the tradeoffs in the efficacy of deterrence if the victim of an attack takes significant time to measure the damage, consult, review options, and most importantly to increase the confidence that attribution of the responsible party is performed correctly?
18. How might international interdependencies affect the willingness of nations to conduct certain kinds of cyberattack on other nations? How can blowback be

exploited as an explicit and deliberate component of a cyberdeterrence strategy? How can the relevant feedback loops be made obvious to a potential attacker?

19. What considerations determine the appropriate mode(s) of response (cyber, political, economic, traditional military) to any given cyberattack that calls for a response?

20. How should an ostensibly neutral nation be treated if cyberattacks emanate from its territory and that nation is unable or unwilling to stop those attacks?

21. Numerous cyberattacks on the United States and its allies have already occurred, most at a relatively low level of significance. To what extent has the lack of a public offensive response undermined the credibility of any future U.S. deterrence policy regarding cyberattack? How might credibility be enhanced?

22. How and to what extent, if at all, must the United States be willing to make public its evidence regarding the identity of a cyberattacker if it chooses to respond aggressively?

23. What is the appropriate level of government to make decisions regarding the execution of any particular declaratory or operational policy regarding cyberdeterrence? How, if at all, should this level change depending on the nature of the decision involved?

24. How might cyber operations and capabilities contribute to national military operations at the strategic and tactical levels, particularly in conjunction with other capabilities (e.g., cyberattacks aimed at disabling an opponent's defensive systems might be part of a larger operation), and how might offensive cyber capabilities contribute to the deterrence of conflict more generally?

25. How should operational policy regarding cyberattack be structured to ensure compliance with the laws of armed conflict?

26. How might possible international interdependencies be highlighted and made apparent to potential nation-state attackers?

27. What can be learned from case studies of the operational history of previous cyberintrusions? What are the lessons learned for future conflicts and crises?

28. Technical limitations on attribution are often thought to be the central impediment in holding hostile cyber actors accountable for their actions. How and to what extent would a technology infrastructure designed to support high-confidence attribution contribute to the deterrence of cyberattack and cyber exploitation, make the success of such operations less likely, lower the severity of the impact of an attack or exploitation, and ease reconstitution and recover after an attack? What are the technical and nontechnical barriers to attributing cyber-intrusions? How might these barriers be overcome or addressed in the future?

D. Regimes of Reciprocal/Consensual Limitations

29. What regimes of mutual self-restraint might help to establish cyberdeterrence (where regimes are understood to include bilateral or multilateral hard-law treaties, soft-law mechanisms [agreements short of treaty status that do not require ratification], and international organizations such as the International Telecommunication Union, the United Nations, the Internet Engineering Task Force, the Internet Corporation for Assigned Names and Numbers, and so on)?

Given the difficulty of ascertaining the intent of a given cyber action (e.g., attack or exploitation) and the scope and extent of any given actor's cyber capabilities, what is the role of verification in any such regime? What sort of verification measures are possible where agreements regarding cyberattack are concerned?

30. What sort of international norms of behavior might be established among like-minded nations collectively that can help establish cyberdeterrence? What sort of self-restraint might the United States have to commit to in order to elicit self-restraint from others? What might be the impact of such self-restraint on U.S. strategies for cyber conflict? How can a "cyberattack taboo" be developed (perhaps analogous to taboos against the use of biological or nuclear weapons)?

31. How and to what extent, if any, can the potency of passive defense be meaningfully enhanced by establishing supportive agreements and operating norms?

32. How might confidence-building and stability measures (analogous to hotline communications in possible nuclear conflict) contribute to lowering the probability of crises leading to actual conflict?

33. How might agreements regarding nonmilitary dimensions of cyberintrusion support national security goals?

34. How and to what extent, if at all, should the United States be willing to declare some aspects of cyberintrusion off limits to itself? What are the tradeoffs involved in foreswearing offensive operations, either unilaterally or as part of a multilateral (or bilateral) regime?

35. What is an act of war in cyberspace? Under what circumstances can or should a cyberattack be regarded as an act of war.[14] How and to what extent do unique aspects of the cyber realm, such as reversibility of damage done during an attack and the difficulty of attribution, affect this understanding?

36. How and to what extent, if any, does the Convention on Cyber Crime (http://conventions.coe.int/Treaty/EN/Treaties/html/185.htm) provide a model or a foundation for reaching further international agreements that would help to establish cyberdeterrence?

37. How might international and national law best address the issue of patriotic hackers or cyber patriots (or even private sector entities that would like to respond to cyberattacks with cyber exploitations and/or cyberattacks of their own), recognizing that the actions of such parties may greatly complicate the efforts of governments to manage cyber conflict?

E. Cyberdeterrence in a Larger Context

38. How and to what extent, if at all, is an effective international legal regime for dealing with cyber crime a necessary component of a cyberdeterrence strategy?

39. How and to what extent, if at all, is deterrence applicable to cyberattacks on private companies (especially those that manage U.S. critical infrastructure)?

---

[14]The term "act of war" is a colloquial term that does not have a precise international legal definition. The relevant terms from the UN Charter are "use of force," "threat of force," and "armed attack," although it must be recognized that there are no internationally agreed-upon formal definitions for these terms either.

40. How should a U.S. cyberdeterrence strategy relate to broader U.S. national security interests and strategy?

F. The Dynamics of Action/Reaction

41. What is the likely impact of U.S. actions and policy regarding the acquisition and use of its own cyberattack capabilities on the courses of action of potential adversaries?
42. How and to what extent, if at all, do efforts to mobilize the United States to adopt a stronger cyberdefensive posture prompt potential adversaries to believe that cyberattack against the United States is a viable and effective means of causing damage?

G. Escalation Dynamics

43. How might conflict in cyberspace escalate from an initial attack? Once cyber conflict has broken out, how can further escalation be deterred?
44. What is the relationship between the onset of cyber conflict and the onset of kinetic conflict? How and under what circumstances might cyberdeterrence contribute, if at all, to the deterrence of kinetic conflict?
45. What safeguards can be constructed against catalytic cyberattack? Can the United States help others with such safeguards?

H. Collateral Issues

46. How and to what extent do economics and law (and regulation) affect efforts to enhance cybersecurity in the private sector? What are the pros and cons of possible solution elements that may involve (among other things) regulation, liability, and standards-setting that could help to change the existing calculus regarding investment strategies and approaches to improve cybersecurity? Analogies from other "protection of the commons" problem domains (e.g., environmental protection) may be helpful.
47. What are the civil liberties implications (e.g., for privacy and free expression) of policy and technical changes aimed at preventing cyberattacks, such as systems of stronger identity management for critical infrastructure? What are the tradeoffs from a U.S. perspective? How would other countries see these tradeoffs?
48. How can the development and execution of a cyberdeterrence policy be coordinated across every element of the executive branch and with Congress? How should the U.S. government be organized to respond to cyber threats? What organizational or procedural changes should be considered, if any? What roles should the new DOD Cyber Command play? How will the DOD and the intelligence community work together in accordance with existing authorities? What new authorities would be needed for effective cooperation?
49. How and to what extent, if any, do private entities (e.g., organized crime, terrorist groups) with significant cyberintrusion capabilities affect any government policy regarding cyberdeterrence? Private entities acting outside government control and private entities acting with at least tacit government approval or support should both be considered.

50. How and to what extent are current legal authorities to conduct cyber operations (attack and exploitation) confused and uncertain? What standards should govern whether or not a given cyber operation takes place? How does today's uncertainty about authority affect the nation's ability to execute any given policy on cyberdeterrence?
51. Cyberattack can be used as a tool for offensive and defensive purposes. How should cyberattacks intended for defensive purposes (e.g., conducted as part of an active defense to neutralize an incoming attack) differ from those intended for offensive purposes (e.g., a strategic cyberattack against the critical infrastructure of an adversary)? What guidelines should structure the former as opposed to the latter?

Research contributions in these areas will have greater value if they can provide concrete analyses of the offensive actors (states, criminal organizations, patriotic hackers, terrorists, and so on), motivations (national security, financial, terrorism), actor capacities and resources, and which targets require protection beyond that afforded by passive defenses and law enforcement (e.g., military and intelligence assets, critical infrastructure, and so on).

2.4 Deterring Cyberattacks: Informing Strategies and Developing Options for U.S. Policy

On June 10–11, 2010, the National Research Council held a workshop entitled "Deterring Cyberattacks: Informing Strategies and Developing Options for U.S. Policy." During this workshop, a number of papers, related to this topic and commissioned by the National Research Council, were presented. These papers were revised and then printed in the published proceedings of this workshop. In addition, the NRC sponsored a prize competition for papers that addressed one or more of the questions raised in Sect. 2.3 above. Two of these papers were singled out for recognition and were included in the published proceedings.

This section contains summaries of the papers published in the proceedings. These summaries were contributed by the authors of those papers. The groupings below reflect the way in which papers were groups in the published proceedings.

2.4.1 Group 1—Attribution and Economics

Introducing the Economics of Cybersecurity: Principles and Policy Options by Tyler Moore

1. Many of the problems plaguing cybersecurity are economic, and modest interventions that align stakeholder incentives and correct market failures can improve our nation's cybersecurity posture substantially.

2. Government should engage Internet service providers (ISPs) in the malware-remediation process by offering exemption from liability for the harm caused by their customers' infected machines in exchange for assisting with the cleanup. The costs of cleanup should be split between ISPs, software firms, and the government, and infection reports should be published on data.gov to encourage better measurement and accounting of these harms.
3. Better data on security incidents are needed to motivate optimal private sector cybersecurity investment. To that end, aggregated reports of online-banking incidents and losses from banks should be collected and aggregated statistics published on data.gov.

Untangling Attribution by David Clark and Susan Landau

1. The occasions when attribution at the level of an individual person is useful are few.
2. Attribution of multistage attacks requires tracing a chain of attribution across several machines.
3. Multistage attacks pose a prime problem for the research community. They should be of central attention to network researchers rather than, for example, the problem of designing highly robust top-down identity schemes.
4. Internet protocol (IP) addresses are more useful than is sometimes thought as a basis of various kinds of attribution.

A Survey of Challenges in Attribution by W. Earl Boebert

1. The Internet has intrinsic features and extrinsic services that support anonymity and inhibit forensic attribution of cyberattacks, and this situation is expected to worsen.
2. Even if perfect forensic attribution were achieved, it would not have a substantial deterrent effect in most cases in which serious disruptive cyberattacks are contemplated by parties hostile to the United States.
3. Alternatives to forensic attribution include counterattack ("hack-back") and sustained, aggressive covert intelligence-gathering and subversion of potential attackers. Such methods promise a greater deterrent effect than forensic attribution. The obstacles to them are primarily nontechnical.

2.4.2 Group 2—Strategy, Policy, and Doctrine

Applicability of Traditional Detserrence Concepts and Theory to the Cyber Realm by Patrick M. Morgan

1. We are fortunate in still being in the early stages of devising responses to the cyber attack threat, and it is hoped that this means that we will avoid the mistakes of

our frantic early—Cold War responses to the Soviet bloc threat—including excessive development of our nuclear-weapons arsenal, the adoption of an unsupportable basic nuclear strategy, and excessive readiness to use nuclear weapons early in any conflict.

2. We must be particularly concerned about the possibility of a strategic surprise first-strike cyber attack in the long run. It is unclear whether such capabilities in cyberspace will ever be developed (they might be), and such an attack would be extremely difficult to detect in advance, so sensitivity to the possibility of one would lead to all sorts of high-alert, potentially overreactive postures on the part of the United States (and a possible opponent)—the worst situation for keeping deterrence stable, as was noted in the study of deterrence during the Cold War.

3. International cooperation to deal with the cyber attack threat is needed because cyberspace is a transnational resource, intended not to be threatening but to be helpful and liberating. It creates a high level of interdependence, so threats that emerge from it must be approached in a multilateral, cooperative fashion. It involves a greater degree of interdependence than that experienced by the antagonists during the Cold War, which led them to develop many elaborate arms-control measures. It is like the interdependence and international cooperation that are now being used or pursued to deal with international terrorism, global warming, and international epidemics and thus is well within our capacities.

Categorizing and Understanding Offensive Cyber Capabilities and Their Use by Gregory Rattray and Jason Healey

1. Offensive cyberoperations can be characterized in many ways. For example, they may be overt, covert, or somewhere in between; or the attacker and the defender or neither can be national military or can be a group with many different kinds of relationships.

2. Many (perhaps even most) of the forms of offensive operations have yet to be seen, and the future of conflict in cyberspace is likely to be very different from the past.

3. The battles of the cyber future may not be "cyber Pearl Harbors" or "digital 9/11s" but may be more analogous to a force-on-force Battle of Britain, a massive support to kinetic operations like the Battle of St. Mihiel, or a long, hard slog over years like the war in Vietnam.

A Framework for Thinking About Cyber Conflict and Cyber Deterrence with Possible Declatory Policies for These Domains by Stephen J. Lukasik

1. A set of long-range security goals suggest 11 unilateral U.S. declarations to initiate processes for the protection of the cybercommons. The declarations are based on the accepted structure of sovereign states as the mechanism to propagate the objectives through eventual international agreements.

2. The declarations assign to each sovereign jurisdiction the responsibility for eliminating the distribution of malware and the capturing of computers for use as botnets in it and the responsibility for attaching a state label to each packet leaving it.

3. The declarations attach to each state that allows harmful packets to leave it potential complicity for any harm suffered by a recipient of the packets. It calls for adjudication of disputes arising from such allegations with appropriate international mechanisms recognized by the parties to such disputes.
4. The declarations imply that attack attribution need go only deep enough to identify the sovereign entities that allowed harmful packets to leave them. But holding "innocent" transit states complicit requires all states to inspect packets coming into them for potential harm and, by implication, to reject them.

Pulling Punches in Cyberspace by Martin Libicki

1. The laws of war do not map very well into cyberspace, because of the potentially large differences between what operations were intended to do, what they actually do, and what they have been perceived to do.
2. Several of the factors that should persuade a state to pull its punches in cyberspace, such as the difficulty of reconciling operations with a state's narrative, the fear of escalation, and the occasional need to take back an action, apply in the physical world but are strongly influenced by the many ambiguities of cyberspace operations.
3. A sub-rosa response to an attack of uncertain effect and attribution has much to recommend it, but it means abjuring attacks on many types of targets. Reliance on sub-rosa responses can promote a lack of accountability among operators.

2.4.3 Group 3—Law and Regulation

Cyber Operations in International Law: The Use of Force, Collective Security, Self-Defense and Armed Conflicts by Michael N. Schmitt

1. The law governing when a cyber operation is a violation of the prohibition of the use of force in the UN Charter and customary international law is unclear. Thus, policy looms large, especially as one may not be able to predict accurately whether other States will deem a given action a violation.
2. The law governing when a State may respond kinetically in self-defense pursuant to Article 51 of the UN Charter and customary international law is relatively clear: the attack must cause (or be intended to cause) death, injury, or damage to property before such a response is lawful. However, States are unlikely to accept that limit in the face of a cyber operation that does not have such consequences when directed against critical assets. Thus, the law should be expected to evolve as State expectations and attitudes crystallize.
3. The law of armed conflict is generally adequate to handle a cyber operation mounted during hostilities. The major point of contention is whether an attack directed against the civilian population or civilian objects is unlawful if it does not injure or kill civilians or damage civilian property. In the view of the author, such operations are lawful.

Cyber Security and International Cooperation by Abe Sofaer

1. Cyber insecurity is an important and expensive problem that is inherently transnational, adversely affects all users worldwide, and is caused by many major players, including parties inside the United States.
2. No state (or group of like-minded states) will be able to deal effectively with all the major aspects of cyber insecurity through defensive and offensive measures.
3. International cooperation is likely to contribute to enhancing cybersecurity in some but not all aspects of current concern through agreements that avoid attempts to regulate inappropriate areas of concern (espionage and aspects of warfare), that seek objectives and use methods consistent with U.S. political and privacy values, and that maintain current, private, professional standard-setting activity rather than transferring such functions to government officials, national or international.

The Council of Europe Convention on Cybercrime by Michael A. Vatis

1. The Council of Europe's Cybercrime Convention has been an effective tool for fostering international cooperation on investigations involving computers and digital evidence. Because of the Convention, more countries have passed substantive laws addressing cybercrime and improved their cyber investigation capabilities, and parties to the Convention assist each other more rapidly and frequently.
2. The principal shortcomings of the Convention are its narrow membership (mostly European countries and the United States; Russia and China are not parties) and the lack of an enforcement mechanism if a country refuses to lend assistance when requested.
3. The Convention therefore could be made more effective by increasing its membership and by imposing costs of some sort on states that refuse cooperation without a legitimate, credible reason. While getting parties to agree to impose any kind of sanctions on uncooperative states seems unrealistic, public exposure of a state's lack of cooperation might have some salutary effect. Moreover, the U.S. could announce that, in the case of highly damaging attacks, it reserves the right to engage in unilateral self-help (such as cross-border searches of computers, or perhaps even counter-attacks on computers responsible for the attacks on computers in the U.S.) when the country from which the attacks appear to be emanating refuses to cooperate and provides no legitimate, credible reason.

2.4.4 Group 4—Psychology

Decision Making Under Uncertainty by Rose McDermott

1. Psychological factors are a critical part of understanding the perception of threat, and the kinds of systematic biases that can influence decision makers when they contemplate how to respond.

2. Overconfidence presents a pervasive and endemic problem for decision makers with regard to attribution in particular.
3. The anonymous nature of cyberspace and the speed with which processes of social contagion can spread information like a virus highlights the fact that deterrence no longer offers a viable strategic response for the uncertainty which characterizes this domain; rather, analogies drawn from the spread of infectious disease provides a more helpful model in thinking about designing more effective response strategies.

2.4.5 Group 5—Organization of Government

The Organization of the United States Government and Private Sector for Achieving Cyber Deterrence by Paul Rosenzweig

1. The potential U.S. government responses to a cyber incident span the whole of government and are not limited to cyber responses.
2. Private sector cybersecurity suffers from the "tragedy of the commons," so some form of collective response is essential.
3. Global supply chain security is weak, and a substantial threat from hardware intrusions has yet to be systematically addressed.
4. Policy makers should consider formalizing public–private cybersecurity cooperation through a publicly chartered nonprofit government corporation akin to the American Red Cross.

2.4.6 Group 6—Privacy and Civil Liberties

Civil Liberties and Privacy Implications of Policies to Prevent Cyberattacks by Robert Gellman

1. The civil liberties and privacy implications of potential policies and processes to prevent cyber attacks raise a host of unbounded, complex, difficult, and contested legal and constitutional issues.
2. Cyber-attack prevention activities will at times make use of the surveillance authority given to the federal government, and the law of surveillance is famously complex. One particularly important element is the absence of a constitutionally recognized expectation of privacy in a person's records held by a third party. The growing importance of third-party storage on the Internet and the technological obsolescence of many privacy statutes increases the tension between communication privacy and cyber attack prevention activities based on surveillance.
3. Anonymity on the Internet is prized by many Internet users for various reasons. A general constitutional right to anonymity has not been clearly defined, and conflicts are likely to arise between cyber attack prevention activities that attempt to

identify users and the interests of those who seek anonymity for whistleblowing, political, or other purposes.

4. The Privacy Act of 1974, the main information-privacy law applicable to the federal government, implements principles of fair information practice. The act, which applies to intelligence and law enforcement agencies, strikes a balance between competing objectives by allowing a partial exemption for the agencies. Similar exemptions would probably be available for cyber attack prevention activities.

5. Licensing of computer users, computers, or computer software is a possible response to cyber attack prevention needs. The United States has experience in licensing people and equipment in a way that generally balances due process interests of individuals with the government's need to function. However, a governmentally established identification or authorization prerequisite to general Internet access would be controversial. The authority of the federal government under the Commerce Clause (in Article I, Sect. 8 of the U.S. Constitution) is likely to clash with First Amendment interests, with much depending on the specific details of any regulatory scheme.

2.4.7 Group 7—Contributed Papers

Targeting Third Party Collaboration by Geoff Cohen
Note: Cohen's paper was awarded First Prize in the NRC Prize Competition for Cyberdeterrence Research and Scholarship "for original first steps in addressing the problem of third-party contributors to cyberinsecurity."

1. Existing cybercrime against U.S. private and public interests is a more pressing threat than future cyberwar.
2. Successful cyberattacks can only occur with the (possibly unwitting) collaboration of many US-based third-party infrastructure providers, such as ISPs, network operators, certification authorities, hosting providers, name registrars, and private individuals.
3. Law and policy need to be adjusted to encourage or enforce more aggressive monitoring, notification, and resolution of computer security issues, across all third party participants.

Thinking Through Active Defense in Cyberspace by Jay P. Kesan and Carol M. Hayes
Note: Kesan and Hayes' paper was awarded Honorable Mention in the NRC Prize Competition for Cyberdeterrence Research and Scholarship "for raising important issues regarding active defense in cyberspace."

1. Is active defense technologically feasible? Active defense technology exists and has been steadily improving in accuracy, but it may need further improvements before an active defense system can be implemented.

2. When would active defense be appropriate? Given various legal and practical considerations, active defense is probably most suitable as a response to denial-of-service attacks.
3. Who should be in control of active defense? For the purpose of consistency in implementation and to avoid escalation problems, the government should oversee active defense rather than having each firm responsible for making decisions about cyber counterstrikes case by case. Legal concerns and alternatives should also be considered, as should a potential process for an active defense program.
4. How can innocent third parties be protected? Liability rules should be in place to protect oblivious intermediaries whose systems are inadvertently harmed by cyber counterstrikes aimed at an attacker who had compromised the intermediary systems.

References

Cha, A. E., & Nakashima, E. (2010). Google China cyberattack part of vast espionage campaign, experts say. *Washington Post*. 14 January 2010.

Corrin, A. (2010). DISA to establish safe haven outside the Internet. *DefenseSystems.com*. 12 February 2010. Available at http://defensesystems.com/articles/2010/02/12/disa-dmz.aspx?s=ds_170210.

Joint Chiefs of Staff (2004). *The national military strategy of the United States of America*. Available at http://www.strategicstudiesinstitute.army.mil/pdffiles/nms2004.pdf.

Markoff, J., & Kramer, A. E. (2009). U.S. and Russia open arms talks on web security. *New York Times*. 13 December 2009. Available at http://www.nytimes.com/2009/12/13/science/13cyber.html.

National Research Council (2002). In *Cybersecurity today and tomorrow: pay now or pay later*. Washington: The National Academies Press.

National Research Council (2007). In *Toward a safer and more secure cyberspace*. Washington: National Academies Press.

National Research Council (2009). In *Technology, policy, law, and ethics regarding U.S. acquisition and use of cyberattack capabilities*. Washington: National Academies Press.

National Research Council (2010). In *Proceedings of a workshop on deterring cyberattacks: informing strategies and developing options for U.S. policy*. Washington: National Academies Press (material from the NRC is reprinted with permission, courtesy of the National Academies Press, Washington).

Sofaer, A. D., & Goodman, S. E. (2000). *A proposal for an international convention on cyber crime and terrorism*, Center for International Security and Cooperation, Stanford University. August 2000.

The NRC Letter Report for the Committee on Deterring Cyberattacks: Informing Strategies and Developing Options for U.S. Policy (2010). Available at http://www.nap.edu/openbook.php?record_id=12886&page=2.

U.S. Department of Defense (2006). Deterrence operations: joint operating concept, Version 2.0, December 2006. Available at http://www.dtic.mil/futurejointwarfare/concepts/do_joc_v20.doc.

White House (2009). Cyberspace Policy Review. Available at http://www.whitehouse.gov/assets/documents/Cyberspace_Policy_Review_final.pdf.

Part II
New Regulatory Cybersecurity

Chapter 3
Duties of Care on the Internet

Nico van Eijk

Abstract Internet service providers currently find themselves in the spotlight, both in a national and international context, with regard to their relationship both with governments and other private parties. This contribution focuses on duties of care as concerns the relationship between government and Internet service providers. The situation in four countries—the Netherlands, the UK, Germany and France—was researched. The (self-) regulation with respect to five separate themes (Internet security and safety, child pornography, copyright, identity fraud and the trade in stolen goods through Internet platforms) was identified. The conclusions promote more emphasis on a value chain approach and suggest improving the model of notice and take down in order to create more certainty.

3.1 Duties of care

3.1.1 Introduction

In this contribution, based on study commissioned by the Dutch Scientific Research and Documentation Centre, specific forms of duties of care on Internet service providers in the Netherlands, France, Germany and the United Kingdom are analysed.[1]

---

[1]This article is based on a study carried out by Dutch Institute for Information Law (Instituut voor Informatierecht, IViR) and the Leibniz Center for Law. The Study was commissioned by the Dutch Scientific Research and Documentation Centre (Wetenschappelijk Onderzoeks- en Documentatiecentrum, WODC). The original version of the report: Prof. Dr Nico van Eijk (IViR, www.ivir.nl/staff/vaneijk.html/) and Prof. Dr Tom van Engers (Leibniz, www.leibnizcenter.org/information/people/tom-van-engers) in collaboration with Wiebke Abel, Catherine Jasserand and Chris Wiersma, *Moving Towards Balance, A study into duties of care on the Internet*, Amsterdam 2010. Developments—including jurisprudence—after the publication of the original version of study are not extensively discussed/included in this chapter.

N. van Eijk (✉)
Institute for Information Law (IViR), University of Amsterdam, Kloveniersburgwal 48, 1012 CX Amsterdam, The Netherlands
e-mail: N.A.N.M.vanEijk@uva.nl

J. Krüger et al. (eds.), *The Secure Information Society*, DOI 10.1007/978-1-4471-4763-3_3,   57
© Springer-Verlag London 2013

Duties of care primarily concern the relationship between the government and Internet service providers and usually take the form of regulation or co-regulation. Where this is not the case, any forms of self-regulation will also be considered. It should be noted that it is often difficult to draw the line between co-regulation and self-regulation.

The relationship between government and Internet service providers may have consequences for the responsibility and liability of Internet service providers.[2] These (civil-law) aspects are beyond the scope of this article.

Internet service providers are understood to mean market parties engaged in providing access to the Internet to end-users.[3] In terms of telecommunications regulation, the activity in question consists of a 'public telecom service'.[4] In addition, these parties are often active as providers of so-called hosting and caching services.

The countries were selected based on the fact that they represent different policy/regulatory systems or because they are known for interesting developments. The European context is also taken into account.

3.1.2 Themes and E-Commerce Directive

The analysis of duties of care takes place from the perspective of five themes with the idea that in principle they represent the most relevant aspects of the underlying problems.

The first theme relates to breaches of Internet security.[5] What kinds of duties of care are provided for in order to deal with privacy breaches or malware placement? Internet security is already subject to regulation on the basis of the European framework for the communication sector.

The second theme relates to child pornography. Child pornography on the Internet is among the subjects that required attention at an early stage in the development of the online environment; Internet service providers have been closely involved in this aspect.[6]

Copyright is the third theme of the study. The focus is not on copyright as such but on the possible involvement of the Internet service provider when it comes to observing and protecting applicable copyrights.

Identity fraud has been included as the fourth theme, especially because in 2007 the European Commission recommended that identity fraud be considered a crime in its own right (van der Meulen 2006; de Vries et al. 2007).

---

[2]On the issue of liability, see van Hoboken (2009).

[3]See OECD (2010) conceptual framework to be adopted, which is one of our sources for the description of Internet service providers: '... Internet service providers are generally meant to signify Internet access providers, which provide subscribers with a data connection allowing access to the Internet through physical transport infrastructure.'

[4]So-called resellers of services offered by others are outside the scope of this definition.

[5]On security, see for instance Coupez (2010). On security, see for instance Coupez (2010).

[6]About child pornography: Stol et al. (2008).

The last theme relates to the question as to whether Internet service providers play a part in the sale of stolen goods, more particularly with regard to offering these goods via such platforms as auction sites.

The themes partly overlap each other or raise similar issues, for instance with respect to security aspects and applied procedures (such as forms of notice and take down) (Schellekens et al. 2007), or in the field of enforcement.

The themes are not dealt with exhaustively in this study, but they are mainly considered from the central study question, i.e. if there is a regulated relationship between the government and Internet service providers, and if so, what kind of relationship.

Several themes have strong ties with the E-commerce Directive ('Directive on electronic commerce')[7] of 2000, more in particular with respect to Internet service providers. The Directive comprises a system in which three activities are distinguished: 'mere conduit', 'caching' and 'hosting'.[8] Mere conduit (Article 12) consists of the unmodified transfer of, or providing access to, information. Mere conduit thus includes the core activity of Internet service providers, i.e. providing access to the Internet. If they do not make any further selections or changes to the information, the Directive excludes liability for such activity. Nevertheless, a court or an administrative authority may demand that a service provider terminates or prevents an infringement. Caching (Article 13) refers to the temporary but unmodified storage of information. Hosting (Article 14) refers to activities associated with the storage of information provided by a recipient of the service. This includes hosting a website or personal pages. With regard to caching and hosting, it is stipulated in the Directive that liability is avoided when providers remove information after they have obtained actual knowledge (with respect to information that is—evidently—unlawful/illegal, or where appropriate, by an order to that effect). This is also called 'notice and take down'.

In the provisions of the Directive on mere conduit, caching and hosting, nothing is stated about duties of care. Parties acting in conformity with the Directive, however, can claim a limitation of their liability. Yet, if member states opt for prescribing the notice and take down principle as binding, the Directive would not oppose this. Market parties can make notice and take down part of self-regulation. In either situation, there is a duty of care which falls within the scope of this study.

In 2007, the E-commerce Directive was extensively assessed in a report by Verbiest and Spindler (2007). In this study commissioned by the European Commission, various trends are observed, which are also discussed in the current study. The angle adopted in the Spindler/Verbiest report, however, is different and focuses on the liability of intermediaries in a general sense. The report is also part of the

---

[7]Directive 2000/31/EC of the European Parliament and of the Council of 8 June 2000 on certain legal aspects of information society services, in particular electronic commerce, in the Internal Market (Directive on electronic commerce), OJ L 178/1, 17.7.2000.

[8]To a large extent, this system has been derived from the US Digital Millennium Copyright Act (DMCA). For further information, see Elkin-Koren (2006).

background documents of the recent consultation on the possible revision of the E-commerce Directive and its outcome.[9]

3.1.3 Methodology

The literature on the legal and policy-based context of the five themes as well as the involvement of Internet service providers has been analysed. The relevant regulations and/or self-regulation have been inventoried and summarized in country-specific studies.[10]

Because of the highly dynamic nature of the subject matter and its ongoing development, a traditional study of literature was deemed insufficient. Instead, the aim has been to validate the findings of the study of literature and enrich them with local information. To this end, visits were paid to the selected countries, and interviews were conducted with 6 to 8 stakeholders in each country.

Meetings took place with representatives of (interest groups of) Internet service providers, governments, regulatory and supervisory bodies, social organizations and independent experts.

As agreed with the interviewees, the results of the interviews have been kept anonymous. The researchers are responsible for the interpretation of the interviews and the processing method.

3.2 Findings

The regulations of the selected countries—the Netherlands, France, Germany and the United Kingdom—have been inventoried. First of all, the relevant legislation and regulations have been identified. Where specific regulations were lacking, it has been investigated whether any forms of self-regulation and/or co-regulation exist.[11]

3.2.1 Internet Security

By virtue of Article 4 of the Directive on privacy and electronic communications adopted in 2002, providers of publicly available electronic communication services

---

[9]http://ec.europa.eu/internal_market/e-commerce/directive_en.htm   and   http://ec.europa.eu/internal_market/consultations/2010/e-commerce_en.   hhttp://ec.europa.eu/internal_market/consultations/2010/e-commerce_en.htmtm.

[10]These country studies are available as appendices to the original study.

[11]In an appendix to the original study, more extensive country reports are made available. These country-specific studies also include references to relevant parliamentary documents, literature and jurisprudence.

(which include Internet service providers as well) are required to take appropriate technical and organizational measures to safeguard the security of the services provided.[12] If necessary, this should happen in conjunction with the provider of the public communication network on which the service is provided. The measures to be taken should ensure a security level that is proportionate to the state of the technology and the costs of its execution. In the second paragraph of the article, it is stipulated that providers are to inform their subscribers of the special risks of network security breaches. If the risk lies outside the scope of the measures to be taken by the service provider, the latter must inform the users of any possible remedies, including an indication of the expected costs.

Article 4 was recently extended in the context of the revision of the European framework for the communication sector.[13] A new paragraph 1a has been added to the article, imposing obligations on the providers regarding access to personal data, protecting stored or transmitted personal data and introducing a security policy with respect to the processing of personal data. The national authorities need to be able to audit the measures taken and to issue recommendations. In a new third and fourth paragraph, a notification obligation is introduced as to breaches related to personal data. Breaches are to be reported to the competent national authority. When the personal data breach is likely to have adverse effects on the personal data or the privacy of a subscriber or individual, the provider shall also notify the subscriber or individual of the breach. Further rules can be laid down at a national level. In addition, the European Commission can adopt technical implementing measures.

In all countries, the content of Article 4 of the Directive on privacy and electronic communications can be found in the national telecommunication acts. In each instance, reference is made to the importance of the protection of privacy and personal data in electronic communications. However, hardly anything substantial can be found on duties of care. It is clear, however, that Internet service providers are understood to have mainly two duties of care. The first pertains to taking suitable technical and organizational measures to safeguard Internet security. The second pertains to informing the end-users about specific risks and measures that can be taken to minimize these risks, in so far as the Internet service provider does not have the obligation itself to take measures. In most countries, the minimum requirements or best practices have not been defined any further in regulations or jurisprudence.

In the Netherlands, on the initiative of the Independent Post and Telecommunication Authority (*Onafhankelijke Post en Telecommunicatie Autoriteit*, OPTA), a process has been started to put the duties of care as laid down in Article 11.3 of the

---

[12]Directive 2002/58/EC of the European Parliament and of the Council of 12 July 2002 concerning the processing of personal data and the protection of privacy in the electronic communications sector (Directive on privacy and electronic communications or e-privacy directive) OJ L 201/37 (31 July 2002).

[13]Amendments to the Framework Directive and the Universal Service Directive: Directive 2009/136/EC of 25 November 2009, OJ L 337/11 (18 December 2009) ('Citizens' Rights Directive') and Directive 2009/140/EC of 25 November 2009, OJ L 337/37 (18 December 2009) ('Better Regulation Directive').

Telecommunications Act into practice. This has resulted in the analysis of relevant issues for the establishment of policy rules. Currently, only rules on the obligation of informing end-users about certain risks have been formulated.

These policy rules have been laid down in the 'Policies for information providers on Internet security' (*Beleidsregels informatieplicht voor aanbieders over internetveiligheid*). Further consultations with the Dutch Government on rules obliging Internet service providers to take security measures have been planned.

OPTA is working with the Dutch National Police Services Agency (*Korps Landelijke Politiediensten*, KLPD) on the basis of a protocol containing agreements on information exchange. The KLPD can act against security breaches to the extent that the national penal law allows for sanctions related to this. In addition, OPTA has its own powers to impose administrative sanctions. Studies have shown that the Netherlands is a pioneer in Europe concerning various Internet security aspects (Dumortier and Somers 2008).

Many Dutch Internet service providers have entered into a covenant in which the intentions have been laid down for the joint combat against botnets. The exchange of information on the basis of the covenant plays a major role in this. End-users should be helped to clear their computers, before they obtain access to the Internet again.

In the United Kingdom, the Internet Services Providers' Association (ISPA UK) has formulated 'best current practices', specifically for the secure handling of e-mail. This document is not compulsory for the members.

In Germany, a provision in the national telecommunications act deals with the organizational measures required of Internet service providers; the provision focuses on the prevention of interruptions, the effects of external attacks and catastrophes. Here, too, further implementation is left to the stakeholders. In addition, an anti-botnet website has been developed on the initiative of ECO (*Verband der deutschen Internetwirtschaft*—Association of the German Internet Industry) and the federal government, through which Internet service providers play an active role in dealing with reported and detected botnets, by means of a call centre that actively helps to clear the computers of the reporting clients. The costs are partly carried by the government.

In France, the spam issue in particular has led to further government involvement. The 'Signal Spam' help line was set up with the assistance of public authorities in collaboration with professional parties. This initiative is in line with the recommendations of the French Association of Internet Service Providers (AFA) on technical measures against spam.

The French Government has made a proposal for a statutory regulation that will oblige Internet service providers to report certain security breaches with respect to personal data to the French supervisory authority in this field (CNIL—*Commission nationale de l'informatique et des libertés*). This proposal can be regarded as a response to the recently extended Article 4 of the Directive on privacy and electronic communications. In both the Netherlands and France, the government has expressed its intention to make this notification mandatory for other services of the information society, and not only for Internet service providers (e.g. web transactions, financial services).

In the interviews, it was emphasized that further concrete steps towards putting in place the duties of care arising from the (new) European directive framework are necessary. The interviewed parties generally indicated that Internet traffic inspections[14] might be in conflict with privacy legislation and principles regarding the confidentiality of (tele)communication. From a technical perspective, however, there are various possibilities. Additionally, on the basis of agreements with customers, Internet service providers filter information because of viruses and spam. Several parties have expressed their concern about the lack of clarity of the legal framework concerning the admissibility of such methods. There is little transparency as to who is affected by these methods and to what extent.

Botnets are clearly a concern for Internet service providers. In the interviews, this problem was discussed as a separate aspect within the Internet security theme and the legal framework arising from the implementation of Article 4 of the Directive on privacy and electronic communications. Internet service providers may face blacklisting due to botnets, causing certain services, such as e-mail, to be disrupted. Although many public sources with location data on botnets are currently available, it is difficult to catch all of them, and extensive work is required to deal with botnets in this way. Establishing the reliability of the public sources mentioned is also difficult.[15] Quarantine measures for such computers seem to be necessary, but limiting Internet access also has an adverse impact. Furthermore, differences in available resources imply that not all Internet service providers would (like to) act against botnets for their customers.

Risks associated with the use of wireless routers have received special attention. The interviewees were asked if the current duties of care in the field of Internet security also cover this issue. It is clear that besides Internet service providers there are several other market parties supplying wireless routers. These parties are not within the scope of the current telecommunication-related legal framework.

Another question in the interviews was to what extent the effectiveness of the measures taken to implement the obligation to provide information as set out in Article 4 of the Directive on privacy and electronic communications, is being supervised. The question arose whether the national government could play an active role in instructing end-users about the safety and security of the Internet or whether it could at least be more closely involved in ensuring that the information actually reaches the end-users.

With respect to Internet security, the question was asked which public authorities could be entrusted with dealing with security breaches. The answer to the question depends on whether a security breach is a national security issue or not. Besides the national telecommunications regulator, other authorities in the field of privacy and national defence could play a role.

---

[14]By using Deep Packet Inspection (DPI), for instance.

[15]In this context, see van Eeten et al. (2010).

3.2.2 Child Pornography

The fight against child pornography on the Internet is supported to a large extent by the private INHOPE initiative, which was started in 1995 and which is backed by the European Union.[16] INHOPE is an association of national hotlines where child pornography (and related activities, including grooming—i.e. contacting children online with the intention of abusing them sexually online and/or offline) can be reported. After verification, the notification is passed on to the relevant authorities. The INHOPE practice can be considered a form of notice and take down.

Child pornography has been on the European agenda for some time. In the Framework Decision of 22 December 2003, it is stipulated that member states are to take measures against the proliferation of child pornography.[17] A proposal has been published to replace the Framework Decision by a directive.[18] Article 21 of the draft directive provides that member states should take measures to block access to child pornography. This blocking should come with the necessary guarantees.[19] Furthermore, member states are to take measures to remove child pornography from the Internet. As stated in the preamble, blocking is important when the information originates from countries outside the European jurisdiction.

In the field of child abuse, the police authorities in Europe are already collaborating intensively in the CIRCAMP[20] programme, and further cooperation between Europe and the United Stated (where apparently most child pornography is hosted) has been announced.[21] Which form is used for blocking, is left to the member states. Self-regulation by Internet service providers on the basis of codes of conduct is mentioned as an option (besides blocking by order of the judiciary or the police on the basis of possibilities to that effect within the civil and/or penal law). The choices for alternatives are partly based on what is permitted by national regulation.

Even before the adoption of the E-commerce Directive, the theme of child pornography received ample attention. In practice, notice and take down is implemented via a system of hotlines in the context of INHOPE, the European organization in this field. The websites of these hotlines act as the first entry point for notifications. In general, the focus is exclusively on publicly accessible Internet traffic, especially websites. These hotlines play an important role in handling notifications

---

[16]International Association of Internet Hotlines, www.inhope.org.

[17]Council Framework Decision 2004/68/JHA of 22 December 2003 on combating the sexual exploitation of children and child pornography, OJ L 13/44, 20.1.2004.

[18]European Commission, Proposal for a Directive of the European Parliament and of the Council on combating the sexual abuse, sexual exploitation of children and child pornography, repealing Framework Decision 2004/68/JHA, Brussels, 29.3.2010, COM(2010)94 final; see also: European Commission, press release IP/10/379, 29.3.2010 and MEMO/10/107, 29.3.2010.

[19]On blocking, i.e.: Callanan et al. (2009).

[20]Cospol Internet Related Child Abusive Material Project (www.circamp.eu).

[21]For the collaboration between Europe and the United States, see http://www.independent.co.uk/news/media/us-eu-to-launch-programme-against-internet-child-pornography-1941748.html.

of child pornography, with the active cooperation of the police and the judicial authorities, also at an international level. Most of the time, Internet service providers send their notifications directly to these hotlines.

EU-initiatives to make filtering of child pornography obligatory raised serious concerns about proportionality and effectiveness. Original plans were abandoned by the European Parliament in February 2011.[22]

In some countries, codes of conduct have been developed which include recommendations for notice and take down with regard to child pornography.

In the context of the European Framework for Safer Mobile Use, providers of mobile telephony in all countries under study have signed framework agreements, in which access to child pornographic material is discussed as well. In these agreements, the providers acknowledge their duty of care to contribute to the removal of child pornographic content on the Internet.

In the Netherlands, a Notice and Take Down Code of Conduct (*Gedragscode Notice and take down*) has been developed by the NICC (National Infrastructure Cybercrime). The code is administered in the framework of the Internet Security Platform (*Platform Internetveiligheid*), where the government and market parties work together. The code of conduct is a declaration of intent that the major Internet service providers have underwritten. Service providers in general can use the code for developing notice and take down procedures. The code of conduct aims at offering a number of options with respect to the application of notice and take down procedures to illegal content on the Internet. Handling such procedures is mainly the task of the providers themselves. The role of the judicial authorities is not described. The legal basis in the Dutch Penal Code for a notice and take down order by a public prosecutor in a criminal context requires some clarification, especially with respect to guarantees for a sufficient judicial assessment of such an order. A revision of this provision was announced at the time of the implementation of the E-commerce Directive, but so far it has not been completed yet. The lack of such guarantees has been detected in both the literature and in recent case law.

Several parties in the Netherlands, including the Child Pornography Hotline (*Meldpunt Kinderporno*) and the Internet Security Platform (*Platform Internetveiligheid*), originally support plans for filtering Internet traffic for child pornography. These plans have been abandoned based on similar arguments as put forward by the European Parliament.

In the United Kingdom, the non-governmental Internet Watch Foundation (IWF) acts as a hotline for child pornography reports. On the basis of self-regulation the IWF plays a binding role, not only bringing Internet service providers and experts together but also involving educational institutions and the general public in combating child pornography. The IWF not only takes care of assessing child pornography notifications, referring them to (international) criminal investigation authorities, but also generates a blacklist used by a high percentage of Internet service providers in the United Kingdom for blocking child pornography on the Internet. In its code of

---

[22]http://www.europarl.europa.eu/sides/getDoc.do?pubRef=-//EP//TEXT+IM-PRESS+20110131IPR12841+0+DOC+XML+V0//EN&language=EN.

conduct, the association of Internet service providers (ISPA UK) also refers to the role of the IWF.

The *Freiwillige Selbstkontrolle Multimedia-Diensteanbieter* (FSM) is a self-regulation body in Germany. In addition to a hotline, the FSM has a code of conduct for its members, including all major Internet service providers. Under the code, the members are required to play an active role in the fight against child pornography, including an obligation to forward notifications to criminal investigation institutions. It also provides for warning members or expelling them from the organization if they do not comply with the provisions of the code.

A recently adopted act in Germany (*Zugangserschwerungsgesetz*), which obliges Internet service providers to block child pornographic material belonging to a list prepared by the national police authority (*Bundeskriminalamt*), seems to be on its way to being abolished. In this context, the German Government has also drawn up individual contracts with Internet service providers, the content of which is not known. This act and these contracts have met with much resistance due to the major breach of communication confidentiality and their impact on privacy and freedom of expression in general. No initiatives have been taken to actually prepare the intended list, and now the reversal of the act is being considered. This has also been confirmed in the interviews.

In France, a signalling procedure defined by the law is used for certain categories of 'particularly harmful illegal content', including child pornography. Consequently, Internet service providers have the legal obligation to forward notifications of child pornography to the relevant public authorities.

In addition, the French Association of Internet Service Providers (AFA) has developed a code of conduct, which is close to the Dutch Code of Conduct on Notice and Take Down. However, the French code exclusively pertains to certain categories of illegal content, including child pornography.

In France, the co-regulatory platform *Forum des droits sur l'internet* has issued several recommendations on child pornography on the Internet.[23] One of these has led to a legislative proposal that provides for the imposition on Internet access providers of the obligation to filter child pornographic content.

In the interviews, it became clear that Internet service providers are willing to cooperate in combating child pornography, but that they keep a weather eye open for measures reaching too far concerning their own liability, in view of the liability restrictions in the E-commerce Directive. They also worry that the imposition of obligations relating to combating child pornography may lead to the creation of further obligations in other fields (such as copyright).

In general, the interviewees were satisfied with how the INHOPE hotline system is functioning. One of the benefits mentioned is that the requirement to classify the notified material can be delegated to the hotlines. Too much involvement in classification could lead Internet service providers to intervene in a random fashion. This could result in an unnecessarily strictly censored Internet. The same could

---

[23]The Forum was closed down in December 2010, because its funding was terminated (http://www.foruminternet.org/).

happen if more practices were to emerge in addition to the hotlines, especially if so-called blacklists were used.

On the basis of the interviews, active monitoring of Internet traffic for the purpose of finding child pornography does not seem to be applied. According to the majority of interviewees, deep packet inspection is considered a disproportionate measure.

Several stakeholders expressed (serious) doubts about the effectiveness of filtering measures. They also warned that active filtering by Internet service providers could lead to the development of new encryption techniques as well as underground networks for the spread of such techniques, which will be difficult to detect. Interviewees emphasized the importance of good support for the parents for teaching sensible Internet use when raising their children.

Several parties referred to the practice in the United States whereby market participants from the financial sector work together to check transactions in order to combat access to child pornography on the Internet.

3.2.3 Copyright

The regime of the E-commerce Directive was partly implemented to establish the position of parties such as Internet service providers with regard to copyright. Supplementary to this, we can refer to the discussion in the context of the New Regulatory Framework (NRF)[24] for the communication sector about the 'three strikes'—or graduated response—issues.[25] Proposals to assign a specific role to Internet service providers in enforcing copyright (with respect to downloading music, video, e-books and games in particular)[26] eventually have not led to European regulations. It should also be noted that in Article 3a of the Framework Directive,[27] it is stipulated that fundamental rights and freedoms are to be observed by member states when taking measures on access to, or the use of, services and applications by end-users.

Similar to the theme of child pornography, the regulations laid down in the E-commerce Directive are the decisive legal framework for the copyright theme in all countries under study. On the basis of this, the duty of care of Internet service providers only pertains to measures for removal of offending content, in the form of notice and take down procedures in the context of caching and hosting activities.

---

[24]The New Regulatory Framework concerns the existing directives for the communication sector and can be found in two directives: Directive 2009/136/EC of 25 November 2009, OJ L 337/11 (18.12.2009) and Directive 2009/140/EC of 25 November 2009, OJ L 337/37 (18.12.2009).

[25]See also TNO/SEO/IVIR (2009) and van Eijk (2011).

[26]In some countries, e.g. the Netherlands, downloading is not punishable; in other countries it is. See the literature in the previous note.

[27]Directive 2002/21/EC of the European Parliament and of the Council of 7 March 2002 on a common regulatory framework for electronic communications networks and services (Framework Directive), OJ L 108/33 (24.04.2002), amended by Directive 2009/140/EC of 25 November 2009, OJ L 227/37 (18.12.2009).

In the Netherlands, a number of court decisions establishing the liability of certain Internet (service) providers for copyright infringement has given rise to a further discussion on the limits of the duty of care of Internet service providers. These cases (see the country-specific study) were primarily heard in courts of lower instance and were mostly about websites that were not entitled to the status of hosting services and the corresponding liability restrictions contained in the E-commerce Directive. In each case, the involvement in copyright breaches was such that the limited definition of hosting activities in this directive did not apply. In one case, an Internet service provider was ordered by the court in a provisional relief procedure to intervene by denying access to a website holder who had unlawfully facilitated a copyright breach. In the literature, there is much criticism on this decision.

In the Netherlands, the private use exception in the current Copyright Act, on the basis of which copying, including downloading, of copyright-protected material for private purposes is a permitted act, has recently been under discussion at a parliamentary level. Such an exception (where copying for private use also covers downloading) cannot be found in the copyright legislation in the other countries under study. A parliamentary commission in the Netherlands has proposed to delete the current exception with respect to downloading. This discussion also dealt with the question of whether and how Internet service providers can play a part in enforcing the proposed new prohibition. There have been proposals on using techniques for this, with which Internet traffic can be checked structurally on the level of the files transferred, such as deep packet inspection and fingerprinting. According to the commission, it should also be provided for by law that Dutch Internet service providers or hosting providers should keep the customer data of individuals and companies that set up websites via their infrastructure. The Dutch Government has indicated they agree with the work group that there are various problems in the field of copyright that need to be tackled. New regulations might include the abolition of the private use exception and the introduction of enhanced enforcement mechanisms (primarily aimed at commercial and large-scale infringements).

In the United Kingdom, the duty of care of Internet service providers has hitherto been based on the liability restrictions of the E-commerce Directive, as implemented in national legislation. By virtue of the Digital Economy Act, however, which was recently passed, Internet service providers are to forward notifications of rightful claimants to alleged infringers actively. On the basis of the new provisions, the providers also need to keep lists of end-users who have been the subject of such notifications. They also need to make these lists with identifiable data available to rightful claimants to help detect repeated breaches by end-users. The Internet user's identity is not to be disclosed by means of these lists. If forwarding the notifications does not result in putting an end to the infringements, Internet service providers can be obliged to impose technical restrictions on the use of Internet connections.

In Germany, the implementation of the E-commerce Directive is decisive for the duty of care of Internet service providers with regard to the protection of copyright on the Internet. The German regulations implement the provisions of the Directive literally.

In France, the new legislation, known as the HADOPI laws, has introduced new obligations for Internet access providers. These obligations are new in comparison with the existing duties of care arising from the E-commerce Directive regarding mere conduit, caching and hosting activities by Internet service providers.

Due to the end-users' obligation to secure their Internet connection to prevent copyright infringements—an obligation laid down in the French Code of Intellectual Property—Internet service providers must propose efficient technical measures that are suitable to that purpose. Such measures are included in a list prepared by the HADOPI authority (*Haute Autorité pour la Diffusion des Oeuvres et la Protection des Droits sur Internet*), which was set up pursuant to the new legislation. Additionally, Internet service providers must inform end-users in their user agreements about the possible sanctions in case of non-compliance with the afore-mentioned obligation. If the HADOPI authority, together with the judicial authorities, decides to intervene, Internet service providers can be required to send warning e-mails to end-users (stating that the unauthorized use has been detected) or, in the event of ongoing negligence, to cut off Internet connections. If Internet service providers fail to cooperate, they may be subject to a penalty.

The interpretation in French jurisprudence of the duties of care of Internet service providers has focused primarily on the limitation of liability for hosting activities, as defined in the implementing legislation of the E-commerce Directive. Like in the Netherlands, the interpretation is usually made by courts of lower instance—and not confirmed by higher courts.

Many cases concern the actual knowledge of hosting providers about the presence of unlawful material, which is required to establish intervention as an obligation for hosting providers, pursuant to the formulation of the liability restriction. In a few cases, hosting providers received an injunction, on the basis of their duty of care, to prevent any attempt to put the same content on the Internet again after it had been removed from a website for the first time.

It was generally emphasized in the interviews that the measures right owners wish to see are not covered by the liability restrictions of the E-commerce Directive. Internet service providers who are asked to detect and block Internet traffic that is in breach of copyright, run the risk of being held liable themselves. Furthermore, doubts were expressed about the technical feasibility of the detection of infringing material, which is passed on or stored by Internet service providers. Sending warning e-mails upon establishing the infringing nature of certain material was mentioned as an option.

Concerning the HADOPI legislation, interviewed stakeholders expressed many doubts. They warned that such stringent legislation might lead to the development and use of encryption technology for the distribution of copyright-protected material. Then, the use of the same technology could be used to share illegal content. Some emphasized that Internet service providers should not be put in the position to monitor Internet traffic or to contribute to punitive measures against end-users. There is also much doubt about the capacity of Internet service providers and of the judicial authorities to support the active approach of copyright protection prescribed by the HADOPI legislation. Investigating authorities also questioned the proportionality of the measures and pointed to the relationship with other investigating authorities with respect to cybercrime.

Some parties pleaded for considering the Internet a universal service, incompatible with drastic measures by Internet service providers. Plans for legislation similar to the French HADOPI regulations seem to be looked upon with growing reluctance in other countries. Many parties also pleaded for restraint when it comes to adopting HADOPI-like legislation. No experience has been gained yet as regards the effectiveness and applicability of such regulations.

Similar questions were raised in the context of the Digital Economy Act in the UK. Another issue with respect to the regulations in France and the UK is how they relate to the new Article 1, paragraph 3a of the Framework Directive, which stipulates that measures taken by member states regarding end-users' access to, or use of, services and applications through electronic communications networks shall respect the fundamental rights and freedoms of natural persons, as guaranteed by the European Convention for the Protection of Human Rights and Fundamental Freedoms and general principles of Community law. This includes the right to privacy and rules on due process.

3.2.4 Identity Fraud

Identity fraud on the Internet is understood to mean appropriating somebody else's identity with the intention of committing unlawful acts. Definitions may vary, but they all boil down to this. In a communication of 2007, the European Commission notes that identity fraud in itself is not made punishable in all member states. It is stated that it is often easier to prove the criminal offence resulting from the identity theft than to focus on identity theft as such. This does not alter the fact that identity fraud is a violation of, for instance, privacy regulations. A study commissioned by the European Commission into identity fraud in the EU Member States is currently being carried out. This may lead to further regulations in 2012.

Appropriating somebody else's identity in itself has not been made punishable in any of the countries under study. This means that, on the basis of the limitation of liability for Internet traffic as defined in the E-commerce Directive and implemented in all countries, Internet service providers do not have any special duty of care with regard to identity fraud.

The problem of online identity fraud has been related in particular to other service providers on the Internet, such as social networking websites and banks facilitating online transactions. Due to their involvement in Internet activities by means of which identity fraud is committed, these parties cannot appeal to the liability exceptions of the E-commerce Directive.[28]

The importance of a notification obligation for Internet service providers for security breaches involving personal data has recently been under discussion in the Netherlands at a parliamentary level. This might contribute to combating identity fraud on the Internet. This notification obligation is related to the new Article 4 of

---

[28]The fact remains that general liability rules and privacy regulation apply to them.

the Directive on privacy and electronic communications, which includes such an obligation for Internet service providers as discussed in the section on the theme of Internet security.

In the United Kingdom, the Fraud Act 2006 was passed, which includes a general penalization of fraud. This act was drawn up so as to include emerging practices with respect to new technologies as well.

In the German debate, phishing in particular has been discussed as a fraudulent practice on the Internet. Phishing is the practice by which existing websites are copied and a certain reliability of these copies is feigned although the websites are fake. These phishing websites are used to lure users into providing their identifiable data, such as log-in data. The discussion concentrated on whether such practices can be punishable under the current criminal legislation. A number of provisions were referred to that could cover phishing.

In France, it has been proposed to make appropriating somebody else's identity a punishable offence. Additionally, a technical tool (IDéNum) has been developed with which the authenticity of an online claim on somebody's identity can be established.[29] The French Government is the initiator of this tool and has made it available for general use by service providers.

From the interviews it becomes clear that it is complicated to have Internet service providers directly cooperate in combating identity fraud online. As an example, the fight against phishing was discussed. Effective combating by Internet service providers is primarily hampered because fraudulent websites use certain IP addresses only briefly or are hosted abroad. Some Internet service providers have indicated they are willing to take action within these technical limits after notifications of phishing websites, to prevent being blacklisted due to hosting such websites. Other measures are technically difficult to apply, and they conflict with the right to communication confidentiality and rules on privacy protection. Some parties warned against bringing too many subjects under the Internet service providers' responsibility.

In general, Internet service providers were not identified as the parties to be made accountable in this context. Social networking sites, banks and credit card companies have been mentioned as relevant parties. It should be noted that these parties already take initiatives to counter fraud, whether or not in collaboration with the government.

Further education of end-users was mentioned several times as a major element in countering identity fraud and has led in various countries to public campaigns, among other things.

3.2.5 Sale of Stolen Goods

The sale of stolen goods on the Internet, particularly the role of the Internet service provider in this, has been given relatively little attention so far on a European level.

---

[29]http://www.idenoum.com/.

Platform providers, such as auction sites, claim they perform hosting services as described in the E-commerce Directive. Meanwhile, some preliminary questions have been referred to the Court of Justice of the EU. This pertains to the eBay v. L'Oreal case, where the issue is not stolen goods but the sale of articles that breach intellectual property rights. Recently, the European Court of Justice decided in this case that service providers who play an active role cannot rely on the exemption of the E-commerce directive.[30] An active role can consist of actions such as to giving knowledge of, or control over, the data relating to the offers for sale, when providing assistance which entails, in particular, optimizing the presentation of the online offers for sale or promoting those offers. The impact of the case is yet unclear (the national court now has to take a final decision), but it seems justified to conclude the decision will affect the position of intermediaries.

It should be kept in mind that the following paragraphs describe the situation prior to the L'oréal/eBay decision.

The sale of stolen goods is mainly discussed in relation to platforms such as those of—globally operating—eBay, which is dominant in the countries under study. Auction and selling platforms are the most important players in the sale of stolen goods via the Internet. It can be derived from the interviews that beyond these platforms there are few problems of significance—for the scope of this study, that is. The conclusion is that the E-commerce Directive is the legal framework within which the discussion on this theme takes place. The status of the platforms involved is a fundamental issue. As regards the sale on auction and selling platforms of goods that breach intellectual property rights, a varying picture has emerged so far from court cases on different levels. All countries have jurisprudence in this field. In the terms of the E-commerce Directive, there is no unequivocal categorization of these platforms.

In Dutch jurisprudence, the status of auction sites such as eBay and Marktplaats (owned by eBay) has not been defined any further in relation to the E-commerce Directive. In case law, several requests for measures in connection with the sale of goods breaching intellectual property rights have been assessed in the context of general liability legislation. In this context, notice and take down is considered a proportional measure in the light of the care that may be required of these websites. In jurisprudence, preventive filtering of advertisements prior to their placement or the compulsory listing of such details as the advertiser's name, address and place of business are not acknowledged as suitable measures.

In the L'Oréal v. eBay case, the High Court of the United Kingdom ruled in favour of eBay and acquitted this organization from liability for material offered by its users that breaches the trademark right of others.

In Germany, the *Bundesgerichtshof* (Federal Court of Justice) ruled in three different cases that online auction websites, in contrast to Internet service providers and other intermediaries, are directly responsible for offering counterfeit and pirated

---

[30]ECJ 12 July 2011, Case C-324/09 (L'oréal and Others v eBay).

goods (*Störerhaftung*). In addition, this court has developed a preventive remedy for right owners against auction websites. This means that auction websites have a duty of care to prevent future breaches of intellectual property rights by users who have already been considered potential infringers. The court has ruled that the use of filter software can help and that such measures are not disproportionate.

Several courts in France, including a court of higher appeal, have ruled that eBay is to be regarded as a hosting provider and that it is not obligated to perform any preventive investigations into the integrity of the advertisements placed. The Court of First Instance for Commercial Law (*Tribunal de commerce*), however, refused to qualify eBay as a hosting provider in three decisions in 2008. This court held eBay liable for its lack of supervision and its failure to take efficient and suitable measures against the sale of counterfeit and pirated goods.

In France, there are several recommendation documents, prepared by expert groups and initiated by the government. One of these pertains to the trade in cultural goods and recommends, among other things, the creation of a register of (stolen) cultural goods. It is specifically aimed at cooperation between online selling platforms and trademark owners to counter the online trade in counterfeit and pirated goods. There have been governmental discussions about which activities of platform providers could be subject to the liability restriction for hosting providers in the E-commerce Directive (and implemented in French law). To date, the recommendations and discussion in this respect have not led to any changes in the legal provisions.

It was widely expressed in the interviews that Internet service providers are not always the proper parties for regulating the online sale of stolen goods. Some of the Internet service providers indicated they had never received a request for intervention with respect to stolen goods. Others indicated they were prepared to cooperate with the judicial authorities and the police if asked to do so. Checking Internet traffic for this aspect is not effective, and it is technically unfeasible. A formal duty of care would lead to excessive intervention by Internet service providers and possibly could escalate in the creation of further duties of care in other fields. Intervention with regard to illegal content in general might be next and would result in disproportionate restrictions on (future) economic activities on the Internet.

In general, platform providers that facilitate the online sale of goods are seen as the key players. These platform providers have introduced self-regulation, on account of the fact that the reactions of the users of such platforms provide a major motivation to take responsibility for this problem. This self-regulation mainly consists of forms of notice and take down procedure by eBay and others, with these parties referring to the liability exception that applies from the implementation of the E-commerce Directive for service providers that perform hosting activities. In their opinion, the exception is also applicable to them.

Platforms for the online sale of goods have taken several initiatives to set up procedures for the handling of complaints about offers of stolen goods and counterfeit and pirated goods. Additionally, users are informed about existing procedures and about the regulations that apply.

There is collaboration with the judicial authorities and the police, who can count on an active response from the platform providers. There are also active consultations with the judicial authorities and the police about the reactions of the original owners of stolen goods. Debate on the sale of stolen goods and fraud often leads to the conclusion that these are civil matters (for instance with regard to claiming compensation for the financial damage incurred).

Intellectual property right holders, especially trademark owners, put much pressure on platform providers. The measures they have asked for, are reflected in several legal proceedings. Their requests have been partly met by the procedure provided via the Verified Right Owner Programme (VeRO), in which eBay has invested in particular.[31] This procedure also relates to the identification of rightful claimants and to identifying relations with advertisements on the platforms afterwards.

3.3 Analysis and Conclusions

The environment of the subject under study is dynamic. In addition to the overview in the previous chapter, some general observations are provided here and conclusions are formulated.

3.3.1 Value Chain

Internet service providers constitute only one of several parties that are active in the value chain between end-users and providers of services (services of the information society as well as other forms of transaction).[32] A provider of an information service uses a hosting provider to make its website accessible on the Internet. Next, the website is opened up via intermediaries, such as search engines and platform providers (auction sites, social networks), before end-users with Internet access via an Internet service provider obtain the information on the website. Another example is that of the end-user who wishes to access an auction/selling site through his Internet service provider to obtain goods that possibly come from a web shop that sells through the auction platform. The operation is handled via a digital bank transaction. Thus, the value chain does not only involve interconnected actions but is also an economic value chain with a multitude of (financial) transactions. Where the role of the Internet service provider could not be determined in the study, it has been investigated whether other intermediaries in the value chain have any duties of care.

Two legal frameworks, both of European origin, play an important role in this context. The Directive on privacy and electronic communications, which is part of

---

[31] http://pages.ebay.nl/vero/.

[32] For this value-chain approach, see Dommering and van Eijk (2010) and Rand Europe (2008).

the directives regulating the communication sector, includes duties of care with respect to Internet security that are relevant for Internet service providers. Secondly, the provisions of the E-commerce Directive need to be taken into account. Although the Directive's rules on 'mere conduit', 'hosting' and 'caching' are focused on the liability of intermediaries, such as Internet service providers, they have also led to duties of care/self-regulation in many countries.

Internet service providers are among the players who are active in the (economic) value chain between end-users and the providers of services. This is confirmed when we hold the five themes up against the light. In several parts, specific duties of care for Internet service providers can be discerned, arising from the sector-specific regulation or in consequence of the rules on E-commerce. With other themes, duties of care are rather seen in relation to other parties in the value chain, more specifically the parties that offer specific services or that facilitate the operation of platforms for such services.

At first sight, putting the responsibility on the Internet service providers seems to be a simple option. After all, the Internet service providers are the ones who control the end-users' access to the Internet. Internet service providers are gatekeepers, and they fulfil a bottleneck job.

At the same time, it becomes clear that this approach is less and less compatible with the dynamics of the Internet (such as the involvement, as described, of many—interacting—parties), with the associated business models, with considerations of efficiency and with aspects of general interest. It is true that Internet service providers are pivotal, but they constitute just one of the parties in a complex value chain. Imposing the duties of care only on the Internet service providers causes an imbalance, which on the one hand does not do justice to the providers' position and on the other hand brings with it some adverse effects for the provision of services and innovation, for instance. After all, Internet service providers will assess their risks on the basis of their own business model. If this allows only a limited risk margin, it is likely that the risks will be ruled out or mitigated, with the result that services that increase the risk will no longer be accessible for end-users or that new services will not be developed. Efficiency considerations are also important: after further testing, seemingly obvious solutions may appear to be inefficient or may appear to lead to high costs (this is the case with filtering or deep packet inspection, for instance). The general interest plays a role when it comes to securing access to the Internet for everybody at affordable rates.

The importance of a value-chain oriented approach is gaining attention in the literature (OECD 2010; Dommering and van Eijk 2010; Rand Europe 2008; Ofcom 2008) but it is also endorsed by many of the interviewees. Internet service providers in particular are critical of the extent to which they are considered to have duties of care. They blame this partly on their high profile and the direct relationship they have with the end-users. At any rate, other parties in the value chain agree that in many cases Internet service providers are not the party with whom the duties of care should rest, and they take a stand themselves as well. This is apparent, for instance, in their involvement in the fight against child pornography, in enforcing copyright, in countering identity fraud or the sale of stolen goods and in promoting Internet security. The concept of a value-chain approach would therefore deserve further attention.

3.3.2 Internet Access/Service Providers

Internet service providers provide access to the Internet to end-users and additionally perform various other tasks, such as hosting personal pages on websites or supplying added value services, such as e-mail. In the study, it becomes clear that sufficient importance should be attached to this distinction. In their capacity as access providers, the Internet service providers are subject to the light E-commerce regime of 'mere conduit' anyway, but they also claim that the message/content is of no concern to them and that they, as transporters, cannot be held responsible for the content of what they transport.

As transporters the Internet service providers are required to respect the confidentiality of communications, it is stated, and therefore they cannot actually bear any responsibility for what Internet users (or service providers) do on the Internet. Some access providers believe that, in principle, they are obliged to allow spam to pass through, for instance—after all, the traffic between providers and users is not to be hampered—but they use spam filters on the basis of a "separate" contractual relationship with the end-users. In this context, it is important to ascertain where the protection that goes with the 'mere conduit' regime of the E-commerce Directive begins and ends. Can the Internet service provider as an access provider be strictly separated from the Internet service provider as a provider of additional services, such as spam filtering? Are such services to be regarded as a separate category or is this a matter of activities that are subject to (or are to be included in) the rules for hosting/caching?

These arguments partly coincide with the viewpoints that are generally expressed in the discussion about net neutrality. Supplementary to this, it is argued that Internet access can be regarded more and more as a universal service. Even though providers are each other's competitors, they believe that end-users are entitled to Internet access and that in principle they cannot discriminate against users at admission.

3.3.3 Notice and Take Down Dominant

In summary, it can be concluded that with three of the five themes (copyright, child pornography and the sale of stolen goods) notice and take down systems are dominant mechanisms. As the occasion arises, the regulations prescribe that Internet service providers are to set up notice and take down procedures to comply with their duties of care. Where no specific legal obligation is in place, the study indicates that Internet service providers have implemented notice and take down procedures at their own initiative so as to be able to appeal to the diminished liability regimen for hosting and caching activities.

However, notice and take down also often occurs outside the circle of parties that are subject to the hosting and caching exceptions, such as among platform providers and other intermediaries (e.g. search engines). They mostly cannot refer to a special legal rule (there are countries that have extended the protection of the e-commerce rules to other players in the value chain, including platform providers) (van Hoboken 2009; European Commission 2003), but they use notice and take down to limit their general responsibility under civil law. Since the legal framework has not been defined any further, it is not clear to what extent a similar appeal to diminished liability is justified, as stipulated for the parties to which the provisions of the E-commerce Directive apply.

Notice and take down procedures have already been the subject of detailed study and evaluation but should be given closer attention.

3.3.4 Local Context

From the stocktaking and analysis of national regulations in combination with the interviews it becomes clear that national circumstances are partly decisive for the way in which the regulations are set up. In the United Kingdom, self-regulation has traditionally been highly developed. This is also reflected in the system adopted for combating child pornography, which goes beyond merely a notification system. In France, the emphasis is rather on regulation through statutory legislation, and self-regulation is clearly less developed than in the United Kingdom. Germany's position is closer to that of the United Kingdom than to the French position. In great outline, the Dutch practice seems to be close to the German position. There is self-regulation, and it works, certainly in the case of child pornography. The code of conduct for notice and take down provides some added value but also has its weak sides, such as the wide possibilities of interpretation and the absence of an enforcement mechanism.

3.3.5 Enforcement

The enforcement of the applicable code faces several critical factors. Firstly, as to enforcement under penal law, there is always a balancing act between the seriousness of the case and the available means. With child pornography, a substantial investigation structure is in place, but it is not always sufficient. Furthermore, where traditional investigation methods—whether or not supplementary—are called in, they appear to be equally effective and at times in themselves sufficient. The associated dilemmas for the Netherlands have already been well identified (Stol et al. 2008), and the interviews show that elsewhere, too, comparable problems are struggled with, including the lack of sufficient knowledge about the technological aspects.

Making filters compulsory was mentioned in several interviews. There is much hesitation about the effectiveness of filtering, which is also confirmed in the literature (Stol et al. 2008; Callanan et al. 2009). Those who really set their minds on it, can easily circumvent the filters. Filters would make things invisible at best, but they do not stop the unlawful activity. Filtering may thus become an excuse for not optimizing the combat against the underlying illegal activities. Other issues are involved as well, however, such as who is liable for the good functioning of filters, what the risks for underblocking/overblocking/mission creep are, what the proportionality of the measures is, etc. These issues are not new, but they always come up in discussions about filtering. Strikingly enough, various respondents (also from the side of the authorities involved) recognize the limits of filtering. Others consider filtering the ultimate remedy: if enforcement comes up against the absence of jurisdiction, filtering could be deployed as an option. Recent developments show that filtering is no longer seen as the ultimate remedy. Initiatives on the European and national level have been abandoned.

In the interviews, it is further indicated that there is much hesitation about deploying criminal measures as part of recent legislation in the field of copyright. Especially in France, where this new legislation is in its implementation stage, there are some doubts as to its effectiveness, for instance with regard to the fact that large groups of the population will be discriminated against and that the regulation has strongly political overtones. Additionally, the social resistance phenomenon is referred to: the authorities involved allegedly have different priorities and would be facing a proportionality problem, and the judicial institutions are said not to have the capacity to deal with a large number of cases. Like elsewhere, the question is asked whose problem is solved here, with an implied reference to the sector's own responsibility as to guarding its own economic interests, such as the development of new business models. Finally, several parties have expressed their concern that peer-to-peer technology will go underground and will use encryption on a massive scale. This would create an untraceable communication network in which large sections of the population participate. There is the risk that this network will also be used for purposes other than merely distributing copyright-protected material.

Deep packet inspection as an enforcement method has been suggested but meets with strong opposition. Internet service providers refer to the principle of confidentiality of communications and state that permanently monitoring all Internet traffic is very expensive. Experts ask questions about the proportionality/legitimacy of deep packet inspection.

When new regulations are imposed, it is important that sufficient attention is paid to the proportionality of the measures proposed and the consequences for enforceability.

3.3.6 Conclusions

A varied picture emerges from the study, which indicates that the developments, including improving the balance within the value chain, are still underway. Internet security, more particularly with regard to the relationship between the Internet service provider and the end-user, is still in its infancy. This does not mean that nothing is happening in practice, but formally a framework has hardly been defined and there is little self-regulation at this stage. On the other hand, there is a virtually identical system for child pornography in the countries under study, where parties are prepared to provide far-reaching assistance in combating this phenomenon. The (INHOPE) notification system is found in all countries either on the basis of self-regulation or in consequence of a legally defined duty of care. The use of filtering is a recurring issue in the prevention of the proliferation of child pornography. Much attention is devoted to copyright, and in two countries the regulations on copyright have been tightened, so that it has become possible to restrict Internet access or to cut end-users off from the Internet. There is strong criticism against the new rules, and from the interviews it becomes clear that the actual enforcement possibilities are subject to much criticism as well. Identity fraud is mainly tackled in the context of the consequences of identity fraud. Making identity fraud punishable in itself (besides the possibilities already in place to act under statutory law) is generally not deemed necessary. The sale of stolen goods via platform providers (e.g. auction and selling sites, etc.) is considered the platform provider's prime responsibility.

The varied picture and the still dynamic nature of the subject make it hard to define proven best practices. Yet, the data gathered in the study provide some interesting information.

On the basis of their study, the following conclusions can be drawn:

1. *Towards a value-chain approach*

Duties of care, as analysed in the study, cannot be linked to one specific party in the value chain between Internet service providers and end-users, but they should be the joint, well-balanced responsibility of the stakeholders in the value chain. Only then, undesired obstacles to Internet access can be prevented and innovation will not be stifled. With the possible introduction of new obligations, it should be assessed in advance what their effects on the value chain will be (such as implications for business models and innovation).

2. *Testing effectiveness and enforceability in advance*

Testing in advance of (intended) legal intervention as regards effectiveness and enforceability contributes to preventing symbolic legislation and undesired (social) effects.[33] What might work in one specific context, might not be the right solution for others due to difference in regulatory and/or judicial traditions.

3. *Deployment of enhanced notice and take down procedures*

Notice and take down procedures appear to be a widely accepted mechanism. The procedures are not only used by Internet service providers (in their capacity as providers of hosting and caching services). Other parties in the value chain, such

---

[33] See the German discussion on filtering of child pornography and what is said in the interviews about the implementation/application of the French HADOPI legislation.

as platform providers, have similar procedures. Most of the countries under study do not have a specific legal basis for these procedures, although there are some initiatives in the field of self-regulation and co-regulation. It is advisable to set a more detailed framework for notice and take down, to define/vary the circle of parties that can use such procedures more closely and to indicate what the effects of such procedures are. Problems related to notice and take down, and more generally the position of the E-commerce Directive, have already been the subject of study but need to be looked into more closely.

4. Clarifying Internet security and privacy

The new rules on Internet security and privacy (Article 4 of the European Directive on privacy and electronic communications) are unclear and require further specification as to their meaning and impact. In principle, it is a European task to prevent differences on a national level that are too significant. A clearer dividing line between security issues that touch on the relationship between Internet service providers/end-users and security issues on a national level is desirable.

5. Increase in the state of knowledge

The need for further regulation is partly fuelled by the lack of sufficient technical and practical knowledge. There appear to be many knowledge gaps in relation to the problems under study in particular. When end-users, supervisors, enforcers and regulators gain further knowledge, this may contribute to less regulation pressure. The importance of education is widely supported.

References

Callanan, C., Gercke, M., de Marco, E., & Dries-Ziekenheiner, H. (2009). *Internet blocking, balancing cybercrime responses in democratic societies*. Research commissioned by the Open Society Institute.

Coupez, F. (2010). Obligation de notification des failles de sécurité: quand l'union européenne voit double... www.juriscom.net. 30 January 2010.

de Vries, U. R. M. Th., Tigchelaar, H., van der Linden, M., & Hol, A. M. (2007). *Identiteitsfraude; een afbakening, een internationale begripsvergelijking en analyse van nationale strafbepalingen*. WODC/Universiteit Utrecht. Identity Fraud; a demarcation, an international comparison of terminology and an analysis of national criminal offences. English Summary: http://www.wodc.nl/onderzoeksdatabase/identiteitsfraude.aspx?cp=44&cs=6796.

Dommering, E. J., & van Eijk, N. A. N. M. (2010). *Convergentie in regulering: reflecties op elektronische communicatie*. Ministry of Economic Affairs, 's-Gravenhage. (Convergence in regulation: Reflections on electronic communications).

Dumortier, J., & Somers, G. (2008). *Study on activities undertaken to address threats that undermine confidence in the information society, such as spam, spyware and malicious software*. Time.lex CVBA, Brussels.

Elkin-Koren, N. (2006). Making technology visible: liability of Internet service providers for peer-to-peer traffic. *9 N.U. J. Legis. & Pub. Pol'y, 15*.

European Commission (2003). First report on the application of directive 2000/31/EC of the European Parliament and of the Council of 8 June 2000 on certain legal aspects of information society services, in particular electronic commerce, in the Internal Market (directive on electronic commerce), COM (2003) 702 def.

OECD (2010). *The economic and social role of Internet intermediaries*, Paris, April 2010.

Ofcom (2008). Ofcom's Response to the Byron Review. http://www.ofcom.org.uk/research/telecoms/reports/byron/.

Rand Europe (2008). *Responding to convergence: different approaches for telecommunication regulators*.

Schellekens, M. H. M., Koops, B. J., & Teepe, W. G. (2007). *Wat niet weg is, is gezien. Een analyse van art. 54a Sr in het licht van een Notice-and-Take-Down-regime*. Tilburg, November 2007.

Stol, W. Ph., Kaspersen, H. W. K., Kerstens, J., Leukfeldt, E. R., & Lodder, A. R. (2008). *Filteren van kinderporno op Internet, Een verkenning van technieken en reguleringen in binnen- en buitenland*. WODC. Filtering child pornography on the Internet an investigation of national and international techniques and regulations. English Summary. http://www.wodc.nl/images/1616_summary_tcm44-117165.pdf.

TNO/SEO/IVIR (2009). *Ups and downs. Economische en culturele gevolgen van file sharing voor muziek, film en games*. A study by TNO information and communication technology, SEO economic research and the Institute for Information Law, commissioned by the Dutch Ministries of Education, Culture and Science, Economic Affairs and Justice. February 2009.

van der Meulen, N. (2006). *The challenge of countering identity theft: recent developments in the United States, the United Kingdom, and the European Union*. Tilburg: International Victimology Institute Tilburg.

van Eeten, M., Bauer, J. M., Asghari, H., Tabatabaiea, S., & Rand, D. (2010). The role of Internet service providers in botnet mitigation: an empirical analysis based on spam data. http://weis2010.econinfosec.org/papers/session4/weis2010_vaneeten.pdf.

van Eijk (2011). File Sharing, note written at the request of the European Parliament's Committee on legal affairs. http://www.ivir.nl/publications/vaneijk/pe432775_en-rev-fin.pdf.

van Hoboken, J. V. J. (2009). Legal space for innovative ordering. On the need to update selection intermediary liability in the EU. *International Journal of Communications Law & Policy, 2009-13*, 1–21.

Verbiest, T., & Spindler, G. (2007). *Study on the liability of Internet intermediaries*. Study commissioned by the European Commission (contract ETD/20-06/IM/E2/69). November 2007.

Chapter 4
The Governance of Network and Information Security in the European Union: The European Public-Private Partnership for Resilience (EP3R)

Kristina Irion

Abstract In public policy information and communications technology (ICT) infrastructures are typically regarded as critical information infrastructures and, thus, require security and protection against cyberthreats. The European Union (EU) Network and Information Security (NIS) policy combines public and private policies at the level of the operators which are highly interdependent. Any NIS policy success rests to an overwhelming degree on the commitment and compliance of the ICT infrastructure operators. Increasingly, policy makers have to pay attention to the supporting governance system which would give best effect to the NIS policy objectives.

This contribution focuses on NIS governance in the EU and explores mechanisms of cooperation between public and private operating ICT infrastructure through the lens of governance theory. It concludes that NIS governance objectives can be pursued in public-private partnerships, but not all functions of NIS policy can be suitably performed at the EU level. Any engagement with the industry needs to be supported by appropriate governance mechanisms that deliver high levels of commitment and compliance by private stakeholders. Against this backdrop this paper critically assesses the European Public-Private Partnership for Resilience (EP3R) and offers recommendations for EU policy makers on a suitable Europe-wide multi-stakeholder governance framework to promote NIS strategy and high-level policy.

4.1 Introduction

The information and communications technology (ICT) sector has long been acknowledged as serving a dual role: First, it is an important sector of economic activity, increasingly contributing to the overall economy and growth. Second, the ICT infrastructure forms the basis for a wide range of activities which are vital for both

Kristina Irion is Assistant Professor at the Departments of Public Policy/Legal Studies and Research Director in Public Policy at the Center for Media and Communications Studies (CMCS) at Central European University.

K. Irion (✉)
Central European University, Budapest, Hungary
e-mail: IrionK@ceu.hu

J. Krüger et al. (eds.), *The Secure Information Society*, DOI 10.1007/978-1-4471-4763-3_4,   83
© Springer-Verlag London 2013

the economy and society (European Commission 2009a, 1; 2009b, 4; OECD 2008, 4, 22). In the European Union (EU), the ICT sector is directly responsible for 5 % of European GDP, with a corresponding market value of €660 billion annually (European Commission 2010, 4). This figure does not yet include the ICT sector's overall contribution to productivity growth which is estimated to amount to 20 percent directly and 30 percent from ICT investments, but its total impact is even wider when considering all ICT enabled economic activities (Ibid.).

Governments around the world have risen to the challenge, and proposed measures that aim to mitigate risk and enhance the resilience of national ICT infrastructures in cooperation with the operators of these infrastructures. The following examples illustrate what is at stake, as well as the limitation of national policy-makers to bring about effective redress when relying on their traditional regulatory toolkit:

- In 2008 and 2010, a submarine communications cable linking Western Europe, the Middle East and South East Asia was damaged in the Mediterranean which affected Internet and telecommunications traffic of the two latter regions to Europe, including alternative routes which carried additional traffic (BBC 2008).
- Cybercriminals use the power of illegal botnets where large numbers of computers can be remotely controlled with the purpose of sending spam or to coordinate denial-of-service attacks. From the largest known botnets, "Mariposa" (in English "Butterfly"), for example, was reported to control between eight to 12 million individual computers at the time it was dismantled by an international team of Internet security companies and nation law enforcement agencies in 2010 (Menn 2010).[1]
- 2010 saw the spread of "Stuxnet," a computer worm of unknown provenance which was designed to infiltrate the Windows operating system and to target industrial equipment by Siemens. The malware was reported to affect the supervisory control and data acquisition (SCADA) systems that control centrifuges, which according to speculations have set-back significantly the Iranian nuclear program (Fieldes 2011).

All three examples have in common that the incident is not limited to one country but causes regional and, as in the cases of "Mariposa" and "Stuxnet," even global, distributed impact. The actual risk scenarios vary, covering online disruption and congestion at the level of ICT infrastructure, to illegal botnets conducting cyber-criminal activities and damaging industrial systems. It serves as an illustration of (1) the technical, logistical and organizational complexity, (2) ICT interconnectedness and interdependencies across sectors, as well as (3) the high degree of uncertainty with regards to the threats, which develop as dynamically as the overall ICT sector. The systemic interdependencies between ICT infrastructures in relation to

---

[1]In the example of illegal botnets, the virtual network of high jacked computers (or "zombies" as they are referred to) can be enlisted for illegal activities against a fee the criminal controlling the botnet levies from their customers.

other sectors can render a local event a transnational cybersecurity incident. In facing these challenges, national public policies need to address this complexity, work in partnership with the stakeholders, and formulate policies that take into account ICT's global ecosystem.

Countries, thus, readily recognize the need for supranational and coordinated approaches to cybersecurity. International policy steering in a variety of intergovernmental fora attempts to diffuse political, technical and economic cybersecurity strategies and best practices at national levels. The United Nations discuss cybersecurity in a politico-military context focusing on cyber-warfare, or in an economic context emphasizing cyber-crime (Maurer 2011, 6). The Council of Europe's "Convention on Cybercrime," which laid the foundation for a common policy for the protection of society against online crime, entered into force in 2004 and has been ratified in 32 nations (including non-member countries). In spite of being a non-binding instrument, the Organization for Economic Cooperation and Development's (OECD) "Guidelines for the Security of Information Systems and Networks" from 2002 has been influential in promoting a culture of Cybersecurity.[2] The International Telecommunications Union (ITU), a specialized United Nations agency, initiated the "Global Cybersecurity Agenda," which is a framework for international cooperation aimed at enhancing confidence and security in the information society.[3]

About a decade ago, the concern about network and information security (NIS) entered EU public policy and immediately ranked high on the policy agenda. For the European Commission (2006a, 3) "networks and information systems are increasingly central to our economies and to the fabric of society" and ensuring the functionality of these systems is of paramount necessity. The European Commission defines NIS as "the ability of a [electronic] network or an information system to resist [...] accidental events or malicious actions that compromise its availability, authenticity, integrity and confidentiality" (European Commission 2001, 9). In 2009, the ever growing dependence on ICT infrastructure led to it being perceived as a critical information infrastructure which catapulted electronic communications and information networks into the leagues of electricity grids, transport networks, health care and water facilities in terms of national security relevance (European Commission 2009a).

It is already conventional wisdom that NIS is beyond what governments can achieve by means of traditional top-down, command and control regulation (Hämmerli and Renda 2010, 85; OECD 2008, 4). Private ownership in public ICT infrastructure and its interconnectedness dictate a multi-stakeholder effort with shared responsibilities (Alderson and Soo Hoo 2004, 1; European Commission 2001, 2;

---

[2] The 2008 OECD "Ministerial Meeting on the Future of the Internet Economy" in Seoul reinforced the attention paid to policies and international cooperation that aim for security and resilience of networked ICT systems (OECD 2008, 7f.).

[3] Also ITU issued extensive guidance on national cybersecurity strategies which is, however, not very conclusive to the issue of regional coherence (see ITU 2011). In 2002, the OECD issued its guidelines for the security of information systems and networks (OECD 2002; see also OECD 2006) which do also not recognize the important dimension of regional policy coherence.

2009a, 2009b, 5; Hämmerli and Renda 2010, 16; Shore et al. 2011, 4). Increasingly, policy makers have to pay attention to the supporting governance system which buttresses NIS policy objectives because it creates the indispensable commitment and compliance on the part of the operators of ICT infrastructure. To this end public-private partnerships (PPPs) for cybersecurity have come into existence at the national level as the prevailing mode of governance (European Commission 2009a, 2009b, 5: OECD 2008, 8; Shore et al. 2011, 4). Also the European Commission (2009a, 2009b, 7) promotes "[a] multi-stakeholder, multi-level approach... taking place at the European level while fully respecting and complementing national responsibilities." A new European Public-Private Partnership for Resilience (EP3R), which was launched in 2009, embodies the ambition to create a European-wide multi-stakeholder governance framework.

This paper analyzes the notion of a European-wide multi-stakeholder governance framework through the lens of governance theory and it reflects critically the prevailing PPP paradigm. Governance theory is most suitable because it conceptualizes the need for new modes of governance that can accommodate an international ecosystem, high (technical) complexity and multi-stakeholder co-operation. Our understanding of governance systems has much advanced and is informed by experiences and observations in various contexts with similar concerns, such as for example environmental policy. This research links to governance network theory, which is underpinned by the literature on incentive-based regulation in order to derive parameters for a successful engagement and proposes measures to better align economic incentives and public policy. A similar approach has been chosen by Dunn-Cavelty and Suter (2009), however, with a focus on national PPPs for cybersecurity. The only study that takes an EU governance perspective is one by the Center for European Policy Studies' (CEPS) Task Force on Protecting Critical Infrastructure in the EU (Hämmerli and Renda 2010). This contribution will take the study further and analyze the governance model underlying EP3R and conclude with recommendations for a European-wide multi-stakeholder governance framework.

The paper is structured as follows: The first section presents the ICT sector's deregulation history and the resulting governance structure of the liberalized ICT sector. The next section offers a concise overview of the challenges of NIS policy, interrogating the roles and incentives of the operators of ICT networks to make investments in security. In section three, the focus is on the NIS policy shaping up at the EU level with a view on governance functions granted to ENISA and the newly set-up EP3R. Governance theory and incentive based regulation are then introduced in order to approach and operationalize the European stakeholder governance challenge. The final section provides an assessment of the European-wide multi-stakeholder governance framework embodied in the EP3R, followed by the conclusions which offer a set of policy recommendations addressed to the EU policy makers on the governance structure supporting NIS policy at the EU level.

4.2 Governing the Liberalized ICT Sector in the EU

Before the 1980s national security had been one of the arguments to justify the telecommunications monopolies prevalent in Europe. Countries would argue that only the state could guarantee the security of the public switched telephony networks (PSTN) and its services (mostly voice telephony and early forms of data communications such as Fax and BTX). When the information society was still a new era to come, the famous Bangemann Report (High-level Group on the Information Society 1994) articulated the need for a regulatory environment absent of exclusive rights which stimulated private investments into ICT infrastructure. The EU legislator followed this recommendation and successively liberalized the telecommunications sector, which was by and large completed in 1998. As one of the accompanying measures intended to compensate for the absence of direct state control, network security and integrity was identified as licensing criteria for the private sector provision of telecommunications infrastructure (European Commission 1994, 28, 31).

Today's ICT sector has dramatically changed from these early days of a one-network paradigm. If we abstract from the often persistent bottlenecks at the level of local fixed infrastructure, ICT markets in the member states have made significant progress in achieving fixed and mobile infrastructure competition. The ICT infrastructure comprises all these privately owned networks, which are for the sake of communications interconnected. The Internet is the paradigmatic example of a network of interconnected networks that spans the globe. Since liberalization, the EU regulates the electronic communications sector, and the regulatory framework is not only concerned with creating a level playing field for competition but also with other public interest objectives, such as universal service, consumer protection and—most relevant to this survey—the security and integrity of networks and services. Member states transpose the regulatory package for electronic communications to their national system of sector-specific regulation, which is then implemented by national regulatory authorities (NRAs). Additionally, the public and the private sector collaborate on information sharing, standard-setting and best practices, testing ICT resilience and business continuity; because it is understood that mandating security is not enough.

Governance at the EU level involves the delegation of competences to new European policy networks in the electronic communications sector and the creation of EU-wide coordination mechanisms. European policy networks are established to foster regulatory harmonization and uniformity of policy implementation across Europe by providing expertise, all the while promoting international regulatory learning. In telecommunications, the European Regulators Group (ERG) was founded in 2002 under EC law, which was replaced as of 2009 with the Body of European Regulators for Electronic Communications (BEREC). This has enhanced competencies for harmonization and is composed of the heads of NRAs of member states (Regulation (EC) No. 1211/2009). The establishment of the European Network and Information Security Agency (ENISA) is a first move towards institutionalized governance at the EU level. The two other relevant collaborations at EU level concern NIS cooperation among member states and European institutions, i.e. the European Forum for Member States (EFMS), and the bespoke European-wide partnership known as EP3R. After the subsequent discussion of the challenges an EU-wide NIS policy faces, the paper returns to the NIS policy framework at the EU level.

4.3 The Network and Information Security Policy Challenge

This section summarizes the literature on the economics of NIS in relation to ICT infrastructure and extrapolates these findings to the EU level. There are two interrelated concerns: First, economic theory implies for various reasons an underprovision of NIS. Second, already at the level of the nation state NIS efforts often lack systemic efficiency and internal consistency, which hampers the overall effectiveness of private initiatives and public policies aimed at improving security and resilience of ICT networks. Both concerns are ascribed in the following sub-sections, however, it is important to bear in mind that the ICT ecosystem is wider and connects many more services and stakeholders and, thus, creates interdependencies that are beyond the scope of this paper (see for example Bauer and Van Eeten 2009; ENISA 2011).

4.3.1 Economics of Network and Information Security

Economic theory which is supported by limited empirical research holds that the optimal level of cybersecurity cannot be achieved by relying on market forces alone. As Andersson and Malm put it:

> All private firms are responsible to their shareholders for operational business risks and have to prepare for contingencies and emergencies. However, in general, market incentives are not compelling enough for private actors to provide the appropriate level of security for society as a whole. To survive in a market-driven economy, companies need to minimize costs and maximize profits. Keeping reserve stock, maintaining redundant systems, and employing back-up staff all cost money. With pressure to cut costs, less resources are available for contingencies and crisis management (Andersson and Malm 2007, 146).

What Hämmerli and Renda (2010, 49f.) refer to as the efficiency-security trade-off certainly occurs with any extensive engagement in NIS that is costly and requires sustained attention; both likely to exceed what customers are willing to pay for (see also Assaf 2008, 11f.; Moore 2010, 5). In their article, Bauer and van Eeten (2009, 710f.) discuss the role of incentives in information security and introduce empirical data on security incentives of players within the ICT value chain, however, excluding ICT network operators. The findings of security-enhancing and security-reducing incentives confirm the existence of efficiency-security trade-off. This trade-off is ascribed to a number of factors which are believed to distort ICT network operators' incentives to invest more in NIS: Market failures, imperfect information and moral hazard, which are now in turn explained.

Market failures are commonly attributed to the public good characteristic of security at large and negative externalities stemming from individual decision-making that impact NIS (Andersson and Malm 2007, 142; Bauer and van Eeten 2009). From a societal perspective, Bauer and van Eeten (2009, 707) then raise the crucial question whether 'the cost and benefits taken into account by market players reflect the social costs and benefits'. With economic theory this question would be denied because the economic incentives of private actors are not aligned to support the societal desirable higher levels of security.

Negative externalities offer a separate economic explanation for market failures that are a result of private sector entities pursuing sub-optimal investments in NIS. According to Andersson and Malm (2007, 143), an externality is an effect of an individual's actions that affects the welfare of others. In the context of cybersecurity, private operators of ICT networks are unlikely to consider the societal effect of a security incident that would disrupt their networks and services beyond what is the operator's individual equation of investments in business continuity and resilience. Consequently, the sum of individual decisions about investments in cybersecurity is unlikely to achieve the societal optimal level of cybersecurity (see also Hämmerli and Renda 2010, 54; Moore 2010, 6).

Imperfect information is cited as another reason why market-based solutions to NIS are likely to be inefficient. When information is incomplete economic actors are not in the position to make informed decisions on risk management (Hämmerli and Renda 2010, 53; Moore 2010, 7). In the context of NIS, the lack of information is particularly pervasive because of ICT's interconnectedness and the related possibility of contamination from other networks, as well as the high level of uncertainty as to the nature of future risks in a highly dynamic technological environment. Individual companies may not be able to shoulder this task on their own, which is why most European national governments facilitate the work of Computer Emergency Response Teams (CERTs), which collect, analyze and disseminate risk-relevant information.[4] Aside from risk information, Andersson and Malm (2007, 144) maintain that '[i]t is costly and extremely difficult to accurately evaluate emergency preparedness'. Nonetheless, the constant assessment of the emergency preparedness and its adequacy in the light of the relevant risk information remains an effort private actors may fall short of implementing, in addition to a similar exercise that would be required on a society-wide basis.

The last explanation as to why there is an underprovision of NIS from a societal perspective is moral hazard. Moral hazard connotes private actors' expectation not to bear the full responsibilities and costs of any large-scale cybersecurity incident because they speculate on government intervention in the event of a major crisis that would effectively bail them out (Andersson and Malm 2007, 144). Other considerations that are bound to limit a private actor's willingness to prepare for large cybersecurity incidents are liabilities that are ultimately capped at the costs of a bankruptcy. Taken together these factors are disincentives for operators to scale

---

[4]For an inventory of CERT activities in Europe see http://www.enisa.europa.eu/activities/cert/background/inv.

up their without doubt existing efforts to ensure the security and integrity of NIS
networks until they have reached a societal optimal level. There is certainly more
that could be invoked to explain this outcome, such as for example rational igno-
rance or behavioral economics (see Hämmerli and Renda 2010, 56f., Moore 2010
5f.), however, for the purpose of this paper it suffices to understand the need to "get
incentives straight" aimed at raising the bar for NIS preparedness.

4.3.2 EU-Wide Policy Coherence

Raising the bar for NIS preparedness alone does not suffice to reach optimal security levels in the interconnected and interdependent ICT sector. Moreover, an overall
effective NIS policy is required that integrates numerous decentralized measures of
various, mainly private, actors. The concept of policy coherence concerns the inter-
play of public policies and individual ICT network operators' NIS measures so that
they are coordinated and reinforce each other.[5] The European Commission recognizes the need for a coherent policy approach that is not limited to the individual
country:

> The high dependence on [critical information infrastructures], their cross-border intercon-
> nectedness and interdependencies with other infrastructures, as well as the vulnerabilities
> and threats they face raise the need to address their security and resilience in a systemic
> perspective as the frontline of defense against failures and attacks (European Commission
> 2009a, 4).

In practice, however, the required coordination between public and private actors,
as well as their partial policies that would bring about system-wide effectiveness is
for various reasons difficult to achieve.

To start with, Andersson and Malm (2007, 145f.) identify a gap between public
and private sector initiatives towards emergency preparedness measures which came
into existence with the privatization of the underlying infrastructures. In a sense, liberalization has disconnected the state's primary responsibility for national security
from the assets which are now privately owned and controlled. This is essentially
not bad since the security of certain critical infrastructures may have even improved
from being a badly managed state asset to becoming a professionally operated private asset. This gap is better perceived as spheres of influence of government and
private actors that do not meet and therefore leave risks unaddressed. NIS policy
is about addressing this gap mostly through a combination of regulation and incentives that would ideally produce an adequate level of risk reliance and emergency
preparedness.

The other challenge to the coherence of NIS policy is the systematic integration
and coordination of all activities and actors at the national, regional and—to some

---

[5]The OECD (2001, 104; 2003, 2) formulated this concept in the context of development policies,
however, its principles are generalizable and can be flexibly adapted to fit other policy areas.

extent—also global level. Each country is responsible for building an organization that supports the coordination among ICT network operators and the public sector into its national approach to NIS before it can achieve any robust level of security. This has led many countries to herald PPPs because it offers a governance structure that would accommodate public and private actors according to their various roles and responsibilities. Setting-up a PPP, however, is not an end in itself, but requires careful management and a buy-in by ICT network operators on the basis of a policy that inasmuch as possible aligns the socially desirable level of NIS with the operators' willingness to invest in NIS measures. In addition, the European Commission argues the rationale for a European-wide integrated approach:

> A purely national approach runs the risk of producing a fragmentation and inefficiency across Europe. Differences in national approaches and the lack of systematic cross-border cooperation substantially reduce the effectiveness of domestic countermeasures, inter alia because, due to the interconnectedness of [critical information infrastructures], a low level of security and resilience of [critical information infrastructures]in a country has the potential to increase vulnerabilities and risks in other ones (European Commission 2009a, 5).

In order to sum up, ICT network operators are certainly willing to take precautionary measures to protect against operational business risks, but they are not compelled to internalize the risks of ICT network disruptions for society at large. The economics of NIS argue for a role of public policy to better align private incentives to enhance the overall levels of NIS preparedness and resilience, but it does not question the competence of the operators of ICT networks to implement NIS measures. Governments are well advised to leave the details of technical implementation of NIS measures to the competent operators, who are better placed to appropriately manage risks posed to the security of their networks and services (see also Hämmerli and Renda 2010, 86, 89). Instead, governments have to devise policies that mitigate known disincentives as well as introduce positive and negative incentives to stimulate appropriate NIS investments by operators (Hämmerli and Renda 2010, 81). Individual measures must be embedded in a governance structure that fosters coordination among public and private actors, here notably the ICT network providers, in the interest of delivering an overall consistent and effective policy at various levels.

4.4 EU Policy for Network and Information Security

The EU's strategy and policy pertaining to NIS has clearly been developed with some priority over the last five years. It should be noted that the EU has no specific competence for NIS as a policy area, which appears to be at first glance more a matter of national security, i.e. a domain reserved for member states. However, the EU has used its powers under Article 95 of the former EC Treaty (now Article 114 of the Treaty on the Functioning of the European Union) to introduce harmonized regulation on the security and integrity of electronic communications networks and services and for the establishment of ENISA. Other EU NIS activities are based on the so-called flexibility clause in Article 308 of the EC Treaty (now Articles 352 and 353 of the Treaty on the Functioning of the European Union):

> If action by the Community should prove necessary to attain, in the course of the operation of the common market, one of the objectives of the Community, and this Treaty has not provided the necessary powers, the Council shall, acting unanimously on a proposal from the Commission and after consulting the European Parliament, take the appropriate measures.

It is beyond the scope of this paper to discuss in depth EU competences in this area, however, it must be observed that measures on the basis of the flexibility clause are passed unanimously by the Council.[6] Hämmerli and Renda (2010, 81) argue in this context for a strict subsidiarity test to be applied "to identify the functions that should exist at EU level and the ones that are most effectively addressed at member state level."

NIS is an umbrella strategy which combines sector-specific regulation, cyber-crime law, and policies aiming at critical information infrastructure protection (CIIP). The EU NIS policy rests on a three-pronged approach (European Commission 2001, 19; 2006a, 3):

(1) Regulatory package for electronic communications;
(2) Cybercrime legislation; and
(3) NIS measures and the CIIP.

In quick succession relevant policy documents have been issued comprising all the instruments available to the EU policy maker. Table 4.1 below offers an overview of all relevant NIS policy initiatives. In the following the European Union's NIS policy and strategy will be summarized in the light of the evolving governance issues.

4.4.1 Regulatory Package on Electronic Communications

The European Union regulatory package on electronic communications contains a number of provisions on the integrity and security of public communication networks. These provisions are addressed to the member states which have the duty to transpose them into their national laws. European Parliament and the Council (2002a) (Article 8 (4)) lists the integrity and security of public communications networks as one of the policy objectives which member states' NRAs have to implement in the interest of the citizens of the EU. European Parliament and the Council (2002b) (Article 23) requires member states to ensure the availability of public telephony in the event of catastrophic network breakdown or in cases of force majeure.

In 2009, amendments to European Parliament and the Council (2002a) introduced a new chapter dedicated to the security and integrity of networks and services. The new regulation assigns responsibilities to operators of electronic communication networks and providers of electronic communications services which required

---

[6]This very provision has long been criticized, however, for undermining national legislative processes because under this provision member states executives adopt EU measures.

Table 4.1 EU activities shaping NIS policy (based on Servida 2010)

Year	Event
2001	Communication on Network and Information Security: Proposal for A European Policy Approach [COM (2001) 298] (European Commission 2001)
2002	Council Resolution on a common approach and specific actions in the area of network and information security [2002/C 43/02]
2003	Council Resolution on a European approach towards a culture of network and information security [2003/C 48/01]
2004	Establishing the European Network and Information Security Agency (ENISA) [Regulation (EC) No. 460/2004] (European Parliament and the Council 2004)
2005	
2006	– Communication on a Strategy for a Secure Information Society—Dialogue, partnership and empowerment [COM (2006) 251] (European Commission 2006a) – Communication on a European programme for critical infrastructure protection [COM (2006) 786] (European Commission 2006b)
2007	Council Resolution on a Strategy for a Secure Information Society in Europe [2007/C 68/01] (Council of the European Union 2007)
2008	– 1st extension of ENISA's mandate [Regulation (EC) No. 1007/2008] – Public consultation on the future of network and information security
2009	– European Commission communication on Critical Information Infrastructure Protection "Protecting Europe from large scale cyber-attacks and disruptions: enhancing preparedness, security and resilience", including CIIP Action Plan [COM (2009) 149] (European Commission 2009b) – Council Resolution on a Collaborative European Approach to NIS [2009/C 321/01] – Presidency Conclusions of the Ministerial Conference on Critical Information Infrastructure Protection, Tallinn (EE) – Update of the regulatory package e-communications, new chapter on security and integrity of networks and services
2010	Adoption of the Digital Agenda for Europe [COM (2010) 245] (European Commission 2010)
2011	– 2nd extension of ENISA's mandate [Regulation (EU) No. 580/2011] – European Forum for Member States issues European principles and guidelines for Internet resilience and stability – Communication on Critical Information Infrastructure Protection "Achievements and next steps: towards global cyber-security" [COM (2011) 163] (European Commission 2011) – Council conclusions on Critical Information Infrastructure Protection "Achievements and next steps: towards global cyber-security"

them to strengthen the resilience of their operations. The regulation mandated the undertaking of appropriate technical and organizational measures commensurate to the risks posed to the security of networks and services (European Parliament and the Council 2002a, Article 13a (1)). This so-called state-of-the-art principle requires risk management that entails particular measures to prevent and mitigate the impact of security incidents on users and interconnected networks (Ibid.). Operators of public communication networks are under an obligation to take appropriate measures to guarantee the integrity of their networks and to ensure service continuity (European Parliament and the Council 2002a, Article 13a (2)).[7]

National regulators and also ENISA are now equipped with new powers to obtain sufficient information from the network operators and service providers about security incidents and to appraise the level of security (European Parliament and the Council 2002a, Article 13a (3), (4), and 14). A new regulatory instrument is the security breach notification. In the event of a breach of security or loss of integrity that manifests itself with some severity, the operators and providers are under an obligation to notify the competent national authorities (European Parliament and the Council 2002a, Article 13a (3)). Where deemed appropriate, the national regulator can pass the information on to other national regulatory authorities and ENISA (European Parliament and the Council 2002a, Article 13a (4)). The regulators will report annually to the European Commission and ENISA on the notifications received and relevant actions taken.

Newly added is the competence of the European Commission to adopt technical implementation measures; however this is not yet relied on (European Parliament and the Council 2002a, Article 13a (4)). This provision is the basis for the introduction of harmonized technical provisions that specify the state-of-the art principle, standards for network integrity and business continuity, as well as measures defining the circumstances, format and procedures applicable for notification requirements. Such technical regulation should be based on European and international standards whenever possible and do not preclude member state actions towards this end. Member states have to transpose the reform of the electronic communications package by 25 May 2011. This reform is a component of the wider NIS strategy in the EU, which has as additional component CIIP.

4.4.2 Policy on Network and Information Security (NIS)

In 2001, the European Commission issued its proposal for a European policy approach to NIS (European Commission 2001). The communication acknowledges the critical function of networks and information systems for a wide array of activities (including for utilities such as water and electricity supply), and that society at large is relying on the security of these systems. The definition of NIS which

---

[7]Privacy in electronic communications also implies the security of communications services. Under the Directive on Privacy and Electronic Communications providers of communications services to the public have to meet specific security obligations which correspond to the general duty of data processors in the European Union's Data Protection Directive concerning the secure processing personal data. Providers are required to safeguard the security of its electronic communications services by taking appropriate technical and organizational measures commensurate with the risks (Article 4 European Parliament and the Council (2002c)); Article 17 (1) of the Data Protection Directive). The 2009 Citizens' Rights Directive, which amended the Directive on Privacy and Electronic Communications, implemented the obligation to draw-up a security policy and to issue a notification in the event of a personal data breach to the users concerned.

features in the introduction of this paper is used until today. The European Commission's NIS policy rationale is threefold:

1. Enhancing the effectiveness of existing legal provisions based on a common understanding of the security issues and the specific means to address them;
2. Formulating policies which would reinforce market processes and increase the effectiveness of the existing regulatory framework; and
3. Responding to the transnational scope of networks and information systems formulating a European Union wide policy (19).

In essence, NIS policy at the European Union level uses a range of reinforcing policy measures, in particular to address the governance deficit through improved communication, coordination and cooperation at various levels and among all actors from the public and the private sectors (European Commission 2005, 2).

A 2006 Communication which aimed to revitalize the European Commission's 2001 proposals carried forward a coherent approach to NIS (European Commission 2006a). It conceives a new strategy for a secure information society "based on a culture of security and founded on dialogue, partnership and empowerment" (3). Central to the outlined strategy is an open and inclusive multi-stakeholder dialogue which reflects the complementary roles of public and private sector organizations in promoting a culture of security. The Council endorsed the development of a comprehensive and dynamic EU-wide NIS strategy and the holistic approach proposed by the Commission (Council of the European Union 2007). In the following, the European Commission consulted with the public on the future of NIS in the EU. The responses back the European Commission's further endeavors to strengthen NIS community throughout the EU, and develop PPP to exchange best practices and enhance the resilience of infrastructures.[8]

As part of a horizontal effort to protect critical infrastructures in the EU the European Commission devised the European Programme for Critical Infrastructure Protection (EPCIP). It addresses critical infrastructures across sectors which, "if disrupted or destroyed, would have a serious impact on the health, safety, security or economic well-being of citizens or the effective functioning of governments in the Member States" (European Commission 2004, 3). It establishes the notion that critical infrastructure protection involves "a consistent, cooperative partnership between the owners and operators of critical infrastructure and Member States authorities" (6). EPCIP follows a sector-by-sector approach which leads to designated policies for the protection of critical information infrastructures. The relevant Directive does not yet identify critical information infrastructures, which is subject to a future review when priority should be given to the ICT sector, Article 3 (3) of the ECI Directive (Council of the European Union 2008). According to a definition, the concept of critical information infrastructures would comprise "ICT systems that are critical infrastructures for themselves or that are essential for the operation of critical

---

[8]The public consultation which had a turn-around of close to 600 contributions is archived at http://ec.europa.eu/information_society/policy/nis/nis_public_consultation/index_en.htm (accessed 12 December 2012).

infrastructures (telecommunications, computers/software, Internet, satellites, etc.)" (European Commission 2005, 19).

The process of EU NIS policy development reached a new dimension when the European Commission published its 2009 Communication entitled "Protecting Europe from large scale cyber-attacks and disruptions: enhancing preparedness, security and resilience" and accompanying European Commission (2009b). It sets out the CIIP Action Plan that argues the need for the Europe-wide multi-stakeholder governance framework and lays out the foundations for an EU Computer Emergency Response Teams (CERT-EU).[9] The subsequent Council Resolution on a collaborative European approach to NIS (Council of the European Union 2009) and the Council conclusions on CIIP entitled "Achievements and next steps: towards global cyber-security" endorse the policy proposals and the progress made (Council of the European Union 2011). The 2010 Digital Agenda for Europe is one of the seven flagship initiatives of the Europe 2020 Strategy. It dedicates several key actions to NIS policy (European Commission 2010), and seeks to exploit and advance the potential of ICTs and to translate this potential into sustainable growth and innovation.

4.4.3 ENISA: Leveraging Cooperation with Expertise

In 2004, the EU made (at first) a measured institutional commitment to NIS when establishing the ENISA which is based in Heraklion, Greece. Under its constituting European Parliament and the Council 2004) (No. 460/2004), ENISA is a European agency; its initial five year mandate has been extended twice to last now until September 2013. ENISA's mission is to enhance the NIS capabilities of European institutions and the member states, in particular the business community thereof. It therefore acts as a hub of expertise in NIS, which encompasses both cybersecurity and the protection of CIIP (Scott et al. 2001, 11).

The objectives and tasks of ENISA were devised with the mission's critical role of facilitating broad cooperation among all NIS stakeholders. In carrying out its specific technical and scientific tasks ENISA has to reach out and connect all relevant actors in public and private sectors, act as a liaison and seek synergies between public and private actors in the member states and at European Union level. As a platform for exchange and cooperation, ENISA conducts consultations, collaborates on risk assessments and management activities, and engages in awareness-raising, the exchange of best practices and acts as a NIS information hub for all users. ENISA's mandate could be therefore described as an attempt to leverage the desired cooperation with expertise.

---

[9] In September 2011, the EU's new Computer Emergency Response Preconfiguration Team (CERT-EU) took up its work. See European Commission's Press Release IP/11/694 of 10 June 2011 "Cyber security: EU prepares to set up Computer Emergency Response Team for EU Institutions." Available at http://europa.eu/rapid/pressReleasesAction.do?reference=IP/11/694 (accessed December 12, 2011).

Pursuant to the Commission's proposal of 2010 to modernize ENISA, it has been strengthened after the body was in a limbo for the last 5 years as a consequence of its short-lived mandate.[10] The initial mandate (which was strictly non-operational) was expanded by assigning breach notification responsibilities to ENISA under Art. 13a and 13b of European Parliament and the Council (2002a) 2002/21/EC as amended by Directive 2009/140/EC. According to the mandate, ENISA would assume increasingly more responsibility regarding EP3R, and after its mandate is reinforced the running of EP3R would be one of its key activities. With a constituency that includes private stakeholders, the agency has built up a good reputation and ability to reach out effectively to the private sector, which would afford it added value in the European context. ENISA also has a role in coordinating European-wide cyber-security exercises, and it contributes to the new CERT-EU, which has been set-up primarily with the aim of enhancing the incident response capabilities of EU institutions and bodies. ENISA will in the near future also be instrumental in the development of a European Information Sharing and Alert System (EISAS) which connects national CERTs with the CERT-EU.

4.4.4 European Public Private Partnership for Resilience (EP3R)

With its launch in 2009, EP3R embodies the Europe-wide governance framework "to involve relevant public and private stakeholders in public policy and strategic decision making discussions to strengthen security and resilience in the context of CIIP" with a European and international dimension (EP3R 2010, 5).[11] EP3R complements the European Forum for Member States (EFMS) which is reserved for public authorities,[12] whereas EP3R serves as the primary venue for exchange and partnership between the public and private sector (Ibid.). Already the impact assessment exercise conducted prior to the CIIP Action Plan in 2009 favors a non-binding and bottom-up approach to CIIP collaboration between public and private actors (2009b). Through a consultative process, public and private sector stakeholders could shape the objectives, principles and structure of this network. However, the foundational Non-paper has borrowed a lot of the language used in earlier European Commission documents, although now endorsed by various stakeholders (EP3R 2010).

According to the understanding expressed in the Non-paper that provides for the establishment of EP3R the high-level objectives are:

---

[10] In the Digital Agenda for Europe, one of the seven flagship projects of the Europe 2020 Strategy, key actions 6 and 28 set out the objective to modernize ENISA (European Commission 2010, 17).

[11] See Non-paper on the Establishment of a European Public-Private Partnership for resilience (EP3R) Version 2.0, 23 June 2010, available at http://ec.europa.eu/information_society/policy/nis/docs/ep3r_workshops/3rd_june2010/2010_06_23_ep3r_nonpaper_v_2_0_final.pdf (accessed 14 December 2011).

[12] The EFMS' achievement are the European principles and guidelines for Internet resilience and stability (European Forum for Member States 2011).

(1) Provide a platform for information sharing and stock taking of good policy and industrial practices in order to foster a common understanding on the economic and market dimensions of security and resilience in the context of CIIP as well as on the roles and responsibilities of public and private stakeholders;

(2) Discuss public policy priorities, objectives and measures with a view to define framework conditions and socio-economic incentives to improve the coherence and coordination of policies for security and resilience in Europe;

(3) Identify and promote the adoption of good baseline practices for security and resilience, with a view to pursue minimum security and resilience standards and coordinated risk assessment approaches (EP3R 2010, 6).

The founding non-paper identifies as core principles complementarity with existing national public-private initiatives, trusted collaboration among stakeholders (in particular when it comes to the sharing of sensitive information by the private sector), a bi-directional value relationship for governments and industry, and finally, an open and inclusive platform for stakeholder contribution as core principles (Ibid.). The aspect of operational information sharing is presently not included in EP3R's mandate because this is part of the ongoing activities of CERTs and PPPs at the national level. EP3R's first three working groups now operationally cover the following areas:

- Key assets, resources and functions for the continuous and secure provision of electronic communications across countries;
- Baseline requirements for the security and resilience of electronic communications;
- Coordination and co-operation needs and mechanisms to prepare for and respond to large-scale disruptions affecting electronic communications (European Commission 2011, 10).

4.5 Critique of Public-Private Partnerships in Network and Information Security Policy

EP3R is the first attempt to introduce a NIS partnership between public and private stakeholders at the EU level. The context that has led to the flourishing of PPPs in many countries, i.e. the distribution of responsibilities between the public and private sectors in the area of NIS prevails also at the EU level: While characterized by private ownership of ICT infrastructure, governments "remain ultimately responsible for defining and leading public policies for the security and resilience" of critical infrastructure protection (European Commission 2009b, 18). In an ideal scenario, societies "approach critical infrastructure protection through a common-good public–private partnership, both sectors working in harmony to achieve a common goal" (Shore et al. 2011, 4).

In practice, however, these ideal partnerships have yet to be actualized and it appears that PPPs are more of a projection of an efficient governance model than that there is empirical evidence that would support this claim at the level of implementation. Andersson and Malm (2007, 140) caution that "such partnerships may instead become a Pandora's box for many governments—an unreliable and unpredictable solution to the problem of under-provision of governance in deregulated sectors of society, particularly in the areas of national emergency preparedness and crisis management". The mounting criticism of the inflationary use of PPPs at national level is summarized by Dunn-Cavelty and Suter:

> The core problems are, first of all, that the term "PPP" can only describe the nature of existing partnerships in a very rudimentary way, and that the majority of so-called PPP in CIP are not really PPP at all; second, that the interests of private business and of the state are often not convergent when it comes to CIP and that PPP are therefore hardly suitable as solutions; and third, that the existing forms of cooperation are too limited (Dunn-Cavelty and Suter 2009, 180f.).

The authors have identified some structural problems with the concept of PPPs that would require "PPPs to exploit synergies in the joint innovative use of resources and in the application of management knowledge, with optimal attainment of the goals of all parties involved, where these goals could not be attained to the same extent without the other parties" (Ibid., 180). The required complementarity of goals may be absent in a PPP that has been set up with the primary object of promoting the national security of ICT network and information systems which according to the prevailing reading of incentives exceed what private sector stakeholders are willing to achieve. They conclude that "[n]early all of the problems that arise where PPP are formed for the purpose of [critical infrastructure protection] can be reduced to the fact that they are primarily intended to enhance security rather than efficiency" (Ibid., 185). The natural tension between efficiency and security in ICT infrastructure has been affirmed earlier which is bound to ultimately impinge on the effectiveness of any PPP if incentives are not attuned to better value security.

That the European Commission is well aware of the critique shows the assessment of national PPPs' actual performance according to which "ownership and implementation by stakeholders appear insufficient," not least because the involvement of the private sector is often inadequate (European Commission 2009a, 6; 2009b, 8). The 2009 impact assessments discuss governance in PPPs in some detail (European Commission 2009b, 8f.). It holds that:

> PPPs are quite challenging to implement in practice, as information exchange mechanisms between governments and the private sector basically become a trust issue. Private companies will share their sensitive information, about critical assets and the problems they have faced, with other stakeholders (including governments) only if such information is treated confidentially (Ibid.).

Nonetheless, this raises the question whether the European Commission consider PPPs an effective instrument at the national level and transplant this assumption to the European level (Ibid.).

Besides overcoming PPPs' known deficits, the pan-European dimension adds additional complexity (ENISA 2010, 7).[13] The official impact assessments recognize the impact of the degree of institutionalization of the process, the nature of information to be exchanged, as well as the incentives to facilitate PPPs, all of which has to be understood in order to build a successful partnership (European Commission 2009b, 8f.). The CEPS Task Force on Protecting Critical Infrastructure in the EU lists several critical factors for the implementation of EP3R:

> In particular, the size of the expected PPP, the need to accommodate several diverging interests at the same table, the sectoral specificities that would have to be merged into a single platform, and the difficulty of allocating responsibility in what is still chiefly a national prerogative may prove very difficult issues to address, and could potentially undermine the success of this very welcome initiative (Hämmerli and Renda 2010, 80).

Now that EP3R has been launched, it is possible to advance this discussion against the background of the Non-paper that provides for the establishment of EP3R (2010) and using governance theory as a nexus of analysis.

4.6 Governance of Critical Information Infrastructure Protection

The security of ICT networks and information systems, within which CIIP policies belong, provides an excellent case study for the transformation from top-down government to new modes of governance nation states have learnt to accept and work with. It ticks all the boxes listed by the literature that would trigger a shift towards the increased reliance on governance in public management. Recent EU CIIP initiatives render the question on the appropriate European-wide governance framework highly relevant. This section shortly revisits theories of governance, networked governance and incentive-based governance in order to interrogate their relevance for CIIP. In a next step, the investigation focuses on CIIP governance in the EU context and the governance framework of EP3R will be assessed in the light of governance theory in order to deduce key aspects of a suitable European-wide governance framework for CIIP.

4.6.1 CIIP Modes of Governance

This section offers an overview of the literature on governance models for CIIP policies. Most literature focuses on the national context and here on the operational

---

[13] ENISA was tasked with investigating barriers and drivers for public and private actors to cooperate in the area of network and information security (see Preparatory Action 2: Identifying drivers, barriers and frameworks for EU sectoral NIS cooperation in the Work Programme (ENISA 2010, 8)), however, it appears this action has been abandoned.

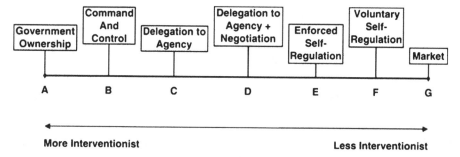

Fig. 4.1 The regulatory continuum of critical infrastructure protection (Reprinted from Assaf 2008, 7[14] with permission)

side of co-operations between public and private actors such as information sharing (Andersson and Malm 2007; Assaf 2008, 2009; Dunn-Cavelty and Suter 2009; ENISA 2010; Shore et al. 2011). National experiences, however, have only limited model character for supranational CIIP governance at the EU level given the different objectives and scope of these policies. Assaf (2008) has conceptualized CIIP models as a regulatory continuum between more and less interventionist governance models as illustrated in Fig. 4.1. His model offers a good entry point to classify the options along a spectrum of intervention intensity, but it falls short of addressing multi-pronged strategies which Bauer and van Eeten (2009, 717) identified as the currently best approach in national efforts to combat cybercrime and enhance information security.

Countries' liberalization and privatization of ICT infrastructure rendered the most interventionist mode A obsolete. As was explained earlier, the merits of mode B, on the one side of the spectrum, which epitomizes command and control regulation are limited because it is likely to be inefficient, inflexible and slow when it comes to the implementation and enforcement of top-down CIIP policies. A purely market based approach as in mode G and even the voluntary self-regulation in mode F, on the other side of the spectrum, are unlikely to produce adequate levels of security given the prevailing economic disincentives for private operators of ICT infrastructure. Bauer and van Eeten (2009, 716) argue for a stronger role of regulatory agencies corresponding to modes C or D because they may be "an efficient intervention point." National regulatory agencies typically have jurisdiction over the critical ICT infrastructure and powers to demand information from the operators are embedded in administrative procedures and sector-specific policy making experience to name just a few (Ibid.). Modes D and E have a co-regulatory component as championed by Dan Assaf who also stresses the need for transparency and public accountability in any such arrangement.

The literature unanimously emphasizes the need for co-operation between public and private sector stakeholders who work in partnership to enhance the security of

---

[14] Assaf's taxonomy suffices for this argument; for an enhanced PPP taxonomy see Shore et al. (2011, 8).

ICT networks and information systems critical for society. PPPs in this area would correspond to modes D to F in Assaf's concept implying different degrees of possible state intervention (Assaf 2009, 68). Already the variations in the definition of PPPs result in a conceptual ambiguity that would prevent them from becoming a point of reference. Observing in their international survey that all countries recognize PPPs' importance, Brunner and Suter (2008, 15) identify different types of such partnerships, such as government-led partnerships, business-led partnerships, and joint public-private initiatives. This epitomizes once again that referring to a PPP is a euphemism which does not resolve the main challenge to identify and implement an effective governance framework. Consequently, literature has nurtured the expectation that governance (network) theory and analyses can provide a concrete recommendations for a multi-stakeholder governance framework (Dunn-Cavelty and Suter 2009, 183; Shore et al. 2011, 6).

4.6.2 Governance (Network) Theory and Analyses

By invoking governance theory this paper explores an, in theory, very successful conceptual framework that seeks to identify governance mechanisms for the joint delivery of a public service by public and private stakeholders. Governance theory took hold in social science at a time when governments are losing their ability to govern exclusively by coercion due to the progressing fragmentation of political power. Causes for this fragmentation are that tasks and authorities are moving beyond the control of central government due to privatization, decentralization and supra-nationalization effects which Rhodes (2000, 71) refers to as the hollowing-out of central government. This coincides with the paradigm shift to new public management that is used to describe a range of state reforms, such as privatization of state functions, aimed at modernizing the public sector towards better management of public resources that emphasizes outcomes and efficiency (Hood 1991, 3). It also links to the internationalization of public concerns that require supranational and joined-up policies beyond what a nation state can realistically achieve on its own.

Hence, according to one of its most influential proponents, Rhodes, "governance is an emergent property of interactions rather than the imposition of control from above" (Pierre and Pieters 2000, 45). Governance is conceptually so amorphous that it can be used to describe the process of governance, the actors and institutions involved in it and the policy instruments used to achieve a particular public policy objective (Rhodes 2000, 55). It revolves around the notion that public and private institutions "are linked by reciprocal connections and more complex network relationships" (Hill 2005, 68). Governments therefore have to use alternative means to shape public policy that may involve re-regulation, soft law but also soft forms of intervention such as network steering, coalition building, moral suasion and networked governance. As a caveat and not surprisingly, governance and governance network theory have both been criticized for their conceptual ambiguity and lack of explanatory power.

Table 4.2 Juxtaposing policy communities and policy networks (based on overviews in Hill 2005, 69)

	Policy communities	Policy networks
Size	Comparatively limited memberships often with economic and professional interests, can be used to exclude others	Large and diverse
Cohesiveness	Shared values and frequent interaction	Fluctuating levels of contacts and comparatively less shared values
Resources	Exchange of resources, with group leaders able to regulate this	Varying resources and an inability to regulate their use on a collective basis
Power	A relative balance of powers amongst members	Unequal power

The idea that governance by networks can considerably enhance public management because it is based on a partnership between public and private sectors that share an interest in a given public service has been very successful (Lane 2009, 64). Lane holds that governance networks' salience must be seen against the background of other popular notions of social capital and trust (Ibid.). Well conceived governance networks are capable of internalizing the knowledge requirement and incentivizing through rewards certain wanted behavior. Other advantages cited in connection with this approach are that participants from different backgrounds collaborating in the delivery of a specific service share their variety of experiences in a framework which levels out hierarchies and compartmentalization (Ibid.; Rhodes 2000, 63). Accordingly, such a setting is believed to motivate participants to perform well and seek out new knowledge and solutions. Thus, networks are likely to "be successful, comparatively speaking, when technology is ill-defined and there is a strong interdependency among the actors at the same time" (Lane 2009, 64).

Networked governance takes place in a variety of possible constellations and PPPs are just one way to refer to networks. In an attempt to come up with a taxonomy of networks the literature discusses "policy networks" (also referred to as "issue networks") and "policy communities" respectively. In both cases the state has a vested interest to foster them, which helps to distinguish them from other purely private interest driven organizations. The differences between policy communities and policy networks are in terms of size, cohesiveness, resources and power which evident from Table 4.2 can be expected to require different governance schemes in order to work effectively for a given model. In addition, the influence on the policy agenda of policy communities is likely to be higher compared to policy networks because of the more homogeneous setting. However, many of these propositions have been derived by observation and there is little that explains why a network develops either way (Hill 2005, 74f.). Another caveat which must be made is that these are not static models, but change can be engendered by changing interests, or from endogenous factors because policy communities and policy networks operate in their specific context.

The theory is becoming less determined when it comes to the management of networks, rewards for engagement and policy implementation in general, but these are the crucial questions that ultimately decide the success of any such approach. In the context of cybersecurity, some authors see an increased reliance on "meta-governance" as the crucial new role of governments, i.e. indirect control as a means to the organization of self-organization (Dunn-Cavelty and Suter 2009, 183; Shore et al. 2011, 6). By the same logic, governments continue to attempt top-down network steering in spite of constraints imposed by networks to exercise authority (Rhodes 2000, 72). Rhodes (2000, 61), however, argues that self-organizing networks tend to resist government steering, which would result in a significant degree of autonomy from the state. Rather, key characteristics of networks are diplomacy, reciprocity and interdependence (Ibid., 61) but to the avail of all network participants. From the point of view of network management, this ultimately carries the risks of increasing the costs of cooperation, the blurring of objectives and suboptimal outcomes. Size too matters from the point of view of effective network management.

Policy network analysis is the attempt to analyze networks' ability to bring about change and to influence public policy making. Several theories concurrently interrogate networks function in terms of participation, agenda setting and actors' behavior but they cannot explain conclusively if, how and why change happens. The study of policy networks is highly circumstantial because its influence is a function of the network itself, its structure and the actors operating in it, all of which is embedded in a given context and a specific policy sub-system. One of the more influential concepts, the Advocacy Coalition Framework by Sabatier and Jenkins-Smith (1993), seeks to explain retroactively over a time perspective of a decade or longer the workings of a network (here: coalition) within a policy subsystems stressing the role of expert information. As a result, neither does policy network analysis offer a forward-looking perspective that could be used as a reference framework to model "successful" governance networks beyond what governance theory above already contributes. Nor does governance (network) theory help eradicate the conceptual ambiguities that have been pointed out in the critique of the PPP model earlier.

4.6.3 Principal-Agent Theory and Incentive-Based Regulation

Borrowing from rational choice theory, the principal-agent framework is another way to conceptualize governance that derives explanations from participants' incentives in a given context. For its representatives "governing involves the manipulation of incentives for the participants, and if those are adjusted properly governing becomes a relative simple exercise" (Pierre and Pieters 2000, 43). However, principal-agent models are becoming less operational and more complex when there are many principals and multiple delegations. As an illustration, public and private sector members in such a partnership represent their organizations and are not automatically enlisted to the objectives for which the partnership has been formed. This is further complicated in the case of a delegation by an EU institution because here too the first instance of delegation has been from the member states to the EU which has then delegated to the agent. By way of incentives it is possible to align the preferences of the partners and their representatives, but any incentive scheme needs to be well conceived so as to produce the desired effects.

In the context of CIIP, Hämmerli and Renda (2010, 74) ask for "the possibility of establishing an effective principal-agent scheme that ties the actions of public and private players to clearly defined objectives, and establishes incentive schemes and sanction mechanisms" (see also Assaf 2009, 74). The details of what would constitute such an effective principal-agent scheme are neither obvious nor easy to conceive. Since precautionary measures and resilience meet uncertainty as regards to the nature of threats and magnitude of risks, a CIIP strategy's only means is the best effort approach. The outcomes of the section on the economics of cybersecurity above suggests a mixed approach that would combine mitigating known disincentives with incentives to stimulate appropriate CIIP investments by operators. For example, a 2010 ENISA study investigates the incentives and barriers to information sharing in which it identifies, but also refutes, certain disincentives to the sharing of security relevant information (ENISA 2010).

Table 4.3 below provides an overview of policy instruments that are proposed in the literature mainly in a national context to incentivize private actors to invest in security, resilience and emergency preparedness of ICT assets. This table distinguishes between relevant positive and negative incentives which are grouped under four categories: legal and regulatory, economic, technical and informational measures. Importantly, their individual effectiveness is not empirically proven and very controversially discussed (Bauer and van Eeten 2009, 715f.; see Hämmerli and Renda 2010, 49f.; Dunn-Cavelty and Suter 2009, 183; Moore 2010, 12f.), which is why there is no ranking among these instruments implied.

Co-regulation, enforced self-regulation or the "shadow of hierarchy" may be necessary to trigger private actors' commitment and compliance. Assaf presents two scenarios from the US (which is traditionally taking a non-interventionist approach), which implies a move from self-regulation to enforced self-regulation with regards to chemical and energy security that has altered the incentive structure of private infrastructure owners to some extent. For New Zealand, Shore et al. (2011, 8) argue the case of enforced self-regulation in CIIP governance. In their survey, Bauer and van Eeten (2009, 716) deduce from an Australian PPP case study on information security that a regulatory threat "seems to have boosted participation" and may have helped that the initiative continued to expand steadily. "Safe harbor" style regulation is another approach to manipulate operators' incentive to contribute to and comply with private CIIP standard-setting in order to benefit from an exemption from a legislative default.

Less invasive is the theory of a "shadow of hierarchy," i.e. the possibility of governmental action, that may be necessary to provoke private policy initiatives such as co- and self-regulation. According to principal-agent-theory, a legislative threat can be perceived as an incentive and some authors would even argue that self-regulation does not exist at all because the motivation is induced endogenously by the possibility of legislative action and private policy making is a strategy to forego regulation.

Table 4.3 Incentives to enhance network and information security of ICT networks (adaptation based on Bauer and van Eeten 2009, 715[a])

Policy instruments	Positive incentives	Negative incentives
Legal and regulatory measures	Public ICT security trustmark	National legislation/regulation of information security
	Setting-up national CERT functionality	Mandating best practices to enhance information security
		Liabilities in case of failure to meet required standards
		Security breach information duties
		Compulsory memberships in professional organizations/PPPs
Economic measures	Tax credits and privileges for certain initiatives	Financial penalties for violations of legal/regulatory provisions (compensatory, punitive)
	Public subsidies for certain investments in information security	Payments for access to valuable information
		Insurance markets
Technical measures	Technical guidance	Information security standards
	Offering technical assistance	Mandating security testing, audits or peer-evaluation
	Education and training relevant to ICT security	Mandating participation in security exercises
Informational measures	National and international information sharing on information security	Publication of individual operator's ICT security breach notifications

[a]Note that Bauer and van Eeten's survey takes the perspective that combines policy instruments to combat cybercrime and promotes enhanced information security of stakeholders in the ICT ecosystem

A different area exhibiting public good characteristics, complexity and interconnectedness is environmental policy which is why research in this field may be relevant to advance our understanding how to achieve private actors' compliance and commitment. Empirical research by Héritier and Eckert (2008) covering a range of environmental initiatives by industry in Germany and the UK shows that in almost all investigated cases voluntary commitments from industry followed a legislative threat. For this investigation it is a relevant insight that in networks, which are said to resist authoritative steering, the "shadow of hierarchy" may be a necessary incentive to enlist the private sector to produce outcomes in the context of a multi-stakeholder policy network. The "shadow of hierarchy" can be direct or indirect, the first being a system of regulated self-regulation, the second referring to a real legislative threat (contrary to the general risk of some legislation to come). Governance network theory and the prospects of introducing a "shadow of hierarchy" are now applied to this investigation into a suitable European-wide governance framework for CIIP.

4.6.4 Implications for the European-Wide Governance Framework for CIIP

Returning to the focus of this paper, the EP3R is just one component of the European-wide governance framework for CIIP that is complemented by ENISA, EFMS, and BEREC. However, as the designated platform for public and private actor collaboration, it is arguably one of the most ambitious initiatives at the European level. As has been previewed, EP3R differs in scope and objectives from its national counterparts (ENISA 2010, 7; European Commission 2009b, 8f.; Hämmerli and Renda 2010, 80). The following assessment relates governance network theory to this partnership, which is then found to conflate a number of concepts discussed in the literature. The argument discusses implications for a European-wide governance framework for CIIP and argues incentive-based governance inspired by principal-agent theory.

In contrast to the comparatively homogeneous policy community, EP3R bears arguably more characteristics of a policy network that is bigger in size and rather diverse. Once EP3R has reached its envisaged constituency it is bound to become rather large, in terms of membership, and diverse as EP3R gives preference to cooperating with the highest ranking executives responsible for NIS from the following organizations: National PPPs and national public authorities in the field of NIS, ICT network infrastructure operators with a European-wide relevance in terms of size or cross-border coverage, and European associations representing ICT infrastructure operators (EP3R 2010, 9f.). Even if the partnership succeeds in attracting only the types of organizations enumerated in the founding non-paper (EP3R 2010, 9), diversity emanates from different national and cultural backgrounds and the relative involvement of public and private sector stakeholders.[15] In the national context, PPP are likely composed by a majority of private stakeholders, but this is not the case with EP3R which reaches out to NIS authorities and national PPPs.

For the desired European-wide governance framework for CIIP this has a number of consequences. There are no regulatory mechanisms at work that would coerce private actors to participate in EP3R. To the contrary, the founding non-paper of EP3R emphasizes a bottom-up approach to CIIP collaboration. At the formal level this organization should preclude direct interventions by EU officials and attempt to exercise indirect control (albeit the European Commission and ENISA facilitate and administer the partnership, EP3R 2010, 9). Governance literature reflects the governability of policy networks which some authors describe as resisting attempts of

---

[15]The organization into different working groups clearly attempts to counterbalance this structural problem and to approximate more the setting of a policy community, but in remains a sub-structure which requires a mandate by the organization and does not produce EP3R-wide consensus.

government (here EU) steering (notably Rhodes 2000, 61). It is important to interrogate the motivations of stakeholders to actively participate and be firmly committed to EP3R's objectives.

At the national level, information sharing has been identified as one of the core motivations in engaging in PPP where the relationship is built on trust among its participants. Trust and credibility cannot be mandated or replaced by binding information frameworks because they are intangible (European Commission 2009b, 34). Diversity and participation can actually become deterrents for sharing sensitive business information (ENISA 2010, 37; see also Dunn-Cavelty and Suter 2009, 182). It is clear that when it comes to creating a trusted environment "[s]upranational PPPs may face a problem of size" (Hämmerli and Renda 2010, 78). In other words, diversity and participation at the European level can actually become deterrents for sharing sensitive business information (7; ENISA 2010, 37; see also Dunn-Cavelty and Suter 2009, 182). Against this backdrop, it is straightforward not to include operational information sharing in EP3R's mandate and refer to its activities being complementary to existing national PPPs information sharing activities.

In fact, EP3R pursues strategic CIIP policy objectives with a view on *best practices*, *statistical frameworks* and *policy recommendations* (EP3R 2010). EP3R's deliberative nature and the interaction with European Commission officials, where the competence for EU policy initiatives rests, may actually explain a fair share of private actors' incentive to participate in EP3R. High-level representation from the European Commission ensures the desired level of executive participation from the private sector and vice versa. EP3R offers an exclusive venue to influence EU policy making at an early stage when it is likely to be most effective. For example, EP3R is involved in discussing a legislative proposal that defines the features that would lead to specific ICT networks being designated critical information infrastructure and trigger regulatory obligations. This and the involvement with similar high-level policy initiatives is what private stakeholders are likely to derive from participating in EP3R.

This leads to the follow-up question whether in the light of its objectives and principles the issues of trust and size are, indeed, such important organizational imperatives for EP3R that they can justify the limitations in representation, accountability and transparency. Since EP3R is effectively a policy network and pursues by and large high-level public policy objectives instead of low-level information exchange, the issues with representativeness and procedural legitimacy are becoming more pronounced. Expressions of interest for participation in the working groups have to be mailed to the European Commission and access to the High Level Steering Group is by invitation only (EP3R 2010, 10). Thus, inclusiveness and participation of EP3R is fairly regulated, which would require a very strong justification in order to counterbalance the inherent legitimacy deficit, which is also a concern of governance theory. Further, it must be noted that apart from the European Commission and ENISA there is no public information about which organizations have endorsed the Non-paper establishing EP3R and that the European Commission's website for EP3R does not feature a list of members in either the High Level Steering Group or the existing three working groups. Since the 2010 non-paper establishing EP3R there is no more recent documentation about its functioning and subsequent activities.

Regulation and governance research, however, stress the importance of principles of procedural legitimacy in co- and self-regulation for the sake of effective and sustainable co-regulation or self-regulation. With regards to PPPs as a governance mechanism to the protection of critical infrastructure protection, Assaf (2009, 78) calls for regulatory arrangements to be restructured in order to enhance the accountability to public values. Transparency crucially accompanies public accountability (Ibid., 77). The 2003 inter-institutional agreement on better law-making (European Parliament et al. 2003), which recognizes alternative regulation mechanisms, among other issues, carries also the commitment that the European Commission:

> will ensure that any use of co-regulation or self-regulation is always consistent with Community law and that it meets the criteria of transparency (in particular the publicizing of agreements) and representativeness of the parties involved. It must also represent added value for the general interest. These mechanisms will not be applicable where fundamental rights or important political options are at stake or in situations where the rules must be applied in a uniform fashion in all Member States (Ibid., para. 17).

Strictly speaking EP3R has not yet made use of co-regulation or self-regulation but it is meant to lead to some sort of commitment of private actors and must adhere to principals of procedural legitimacy. EP3R does not comply with the self-adopted standards of the European institutions in terms of transparency and possibly also representativeness. It clearly infringes aspects of procedural legitimacy whenever EP3R is involved in negotiating important political options of CIIP. Although the EP3R non-paper broadly defines objectives, principles and even expected outcomes, there are no mechanisms that would render the activities of EP3R accountably to the public, let alone to the public good NIS it seeks to promote. Admittedly, any PPP governance arrangement is a delicate maneuver along the spectrum of possible interventions, yet EP3R appears to replicate known deficits of national PPP in the field of network and information security.

However, at the EU level the rationale for setting-up a PPP in the area of NIS follows a path-dependent logic that has its basis in a national PPP, where operational information-sharing requires high levels of trust and confidentially. In the light of EP3Rs objectives and scope of activities, trust and size, which are used to justify limitations to the principles of representativeness, accountability and transparency, are overemphasized. Apart from considerations of procedural legitimacy, the risks are that EP3R will not produce significant and tangible commitments by private operators of ICT infrastructures. The mechanisms that would attune the security-efficiency trade-off and incentive European-wide NIS policy consistency, which is by and large the raison d'être of EU NIS policy intervention, are still at the drawing-board and EP3R is a venue to influence these new regulations to come.

4.7 Conclusions

The security of ICT networks and information systems poses a particular governance challenge for policy makers at all levels. The governance approach is imperative because it is the only means to public policy making under the impression of an international ICT ecosystem that is characterized by high (technical) complexity and requires multi-stakeholder co-operation. In the EU, member states have already accumulated a fair number of years of experience and existing comparative surveys reveal a range of commonalities in the approach and engagement with private sector stakeholders, in particular the creation of PPPs involving the operators of ICT (Brunner and Suter 2008; ENISA 2009).

Aside from raising the overall level of preparedness and resilience against cyberthreats, a central EU NIS policy objective must be regional policy coherence in order to address transborder risks stemming from the ICT sector's interconnectedness. For the EU there is a strong case to set up a dedicated structure with the objective to tackle CIIP in a European-wide context. Aside from the interaction with the private sector ENISA already facilitates, the founding of EP3R in 2010 manifests the first European-wide partnership between public and private stakeholders towards the protection of critical information infrastructure. For the EU, EP3R is the organizational vessel that should provide a Europe-wide governance framework to enhancing cybersecurity.

The national reliance on PPPs as a suitable governance model has clearly informed the EU's NIS governance framework in the area of CIIP (EP3R 2010; European Commission 2009a, 6; 2009b, 8f.). EP3R, however, is in many ways bound to be different from similar initiatives at the national level, where often the same private stakeholders are already engaged. The non-paper establishing EP3R shows acute awareness of the need to observe subsidiarity and complementarity with national initiatives and of the overall context in which this partnership operates (EP3R 2010):

(1) The founding non-paper of EP3R emphasizes a bottom-up approach, the success of which would depend on the active participation and strong commitment of all participants.
(2) EP3R does not engage in operational information sharing and exchange, which is believed to be the main motivation for private actors to engage in national PPPs.
(3) EP3R's high-level objectives centre around "public policy and strategic decision making discussions to strengthen security and resilience in the context of CIIP" (EP3R 2010, 5).
(4) One design principle of EP3R is the creation of a "trusted collaboration" environment (EP3R 2010, 7), which is why participation is limited to NIS senior representatives from enumerated public and private sector organizations.
(5) EP3R pays little attention to procedural legitimacy, such as transparency and accountability to the public and to the public good security.

This paper concludes that EP3R has adopted a restrictive governance structure which can not be entirely justified in the light of its mandate and scope of activities. Deliberations in EP3R are moved up to such crucial tiers in European public policy formulation, that the wish to stay among peers ("trusted collaboration") conflicts with the principles of representativeness, accountability and transparency, contrary to what EU institutions committed to under the 2003 inter-institutional agreement on better law-making (European Parliament et al. 2003). EP3R's selling point is the close engagement with senior representatives from the European Commission and member states' authorities, which appears to be the major incentive for private ICT network operators' motivation to participate. Apart from considerations of procedural legitimacy, the risks are that EP3R will not produce significant and tangible commitments by private operators of ICT infrastructures.

An investigation into a suitable European-wide multi-stakeholder governance framework reveals that PPPs are no silver-bullet to NIS governance (Dunn-Caveltry and Suter 2009, 179) and that the take-aways from governance network theory and analysis are at best mundane. Governance literature does not support the engineering of a successful multi-stakeholder partnership aimed at the protection of the public good NIS, and the study of policy networks implies they tend to resist any attempts of government (here EU) steering where they exist. Thus, governance needs to manipulate the incentives of stakeholders in order for them to carry security preparedness and resilience beyond what efficiency dictates. This paper argues in favor of safe harbor style regulation that would establish a system of enforced self-regulation in CIIP at the EU level.

The tensions created by the need to observe subsidiarity and complementarity with national CIIP initiatives may preclude the adoption of a narrow PPP model for European-wide CIIP governance. The conclusion offers recommendations on an improved European-wide governance framework for CIIP that are informed by procedural legitimacy and incentive-based governance. It does not argue to abandon the notion of a partnership between public and private actors but it tries to untangle this notion from those considerations that are invoked in favor of non-transparent and clandestine arrangements in the context of operational information sharing in national PPPs.

Mixed Policy Approach to Network and Information Security   Governance networks to which PPPs belong are formed with the objective to jointly deliver a public service. NIS is not a public service but a public good with the main difference that economic incentives may produce efficiency but not—from a societal point of view—adequate levels of security. Theory and national practice point towards a mixed approach that would combine mitigating known disincentives with incentives to stimulate appropriate CIIP investments by operators.

At the EU level this result holds true but the mix is different to what is done at the national level, not least because of the need to observe the principles of subsidiarity and complementarity. New regulation of security breach information and the introduction of the state-of-the-art-principle for appropriate technical and organizational measures, the cornerstones of the policy mix, are already in place. A European-wide CIIP governance framework needs to focus primarily on enhancing policy coherence, i.e. the systematic integration and coordination of all activities and actors at regional level.

Not Scaling Trust but the Benefits of Information-Sharing at the EU Level
Trust is not a essential requirement of a European-wide CIIP governance framework

that does not engage primarily in operational information sharing. Rather, the benefits of information-sharing should be scaled to European levels. The key to information sharing is to de-sensitize information from their originating context and repackage CIIP information in a way that would maintain the value of the information for other users. Existing operational information sharing platforms at the international level, e.g. the Forum of Incident Response and Security Teams (FIRST), mark a development towards preserving the benefits of information sharing across borders. Additionally, the EU follows a joined-up approach in which national CERTs exchange information via EISAS and national regulatory authorities in the electronic communications sector share security breach notifications when it is deemed appropriate and an annual summary report on security breach notifications.

European-Wide CIIP Governance Pursues Strategic and High-Level Public Policy Objectives   Different to national PPPs in the field of network and information security, European-wide CIIP governance involving public and private actors pursues strategic and high-level public policy objectives. The deliberations must be open and transparent in order to prevent them from turning into an exclusive lobbying venue for private stakeholders with vested interests in CIIP policies. Instead CIIP governance requires sustained attention at the EU level and an annual conference could be a suitable venue for public deliberation. In order to collect sector-specific expertise, public consultations on EU policy and legislative proposals may be a suitable and procedurally legitimate alternative.

European-Wide CIIP Governance Has a Large and Diverse Constituency
A European-wide engagement with public and private actors results inevitably in a large and diverse constituency. Any efforts to contain membership are most unlikely to make a difference on the commitment of the participants. Since in European-wide CIIP governance trust and confidentiality are less of a prerequisite, participation must be open to all stakeholders including representatives, who are not representing industry such as civil society organizations.

Co-Regulation of Private CIIP Standard-Setting Activities Likely to Be Most Effective   Co-regulation or enforced self-regulation is likely to be the most effective CIIP governance mechanism to correct the underprovision of network and information security. It combines the advantages of industry CIIP standard-setting, i.e. expertise, flexibility, compliance and monitoring, with EU-wide regulatory benchmarks geared towards efficient strategies, policy coherence and accountability. The EU policy maker should consider regulation that would trigger "safe harbor"-privileges for operators of ICT networks which adhere to industry best practices and which have been endorsed by the competent authorities.

Effective CIIP Policy Cooperation Needs Transparency and Accountability
An effective CIIP policy cooperation between public and private actors at the European level needs transparency and accountability. According the 2003 inter-institutional agreement on better law-making, this would be already mandatory because the European-wide CIIP governance classifies as an alternative regulation

mechanism. Beyond formal obligations, European-wide CIIP governance is best served by public scrutiny and accountability to the public interest, which helps to assess whether deliberations on high-level policy are not unduly influenced and private actors' commitments hold strong.

At the EU level, efforts should be made to correct the EP3R governance model following these recommendations. The corrected European-wide multi-stakeholder governance framework would become susceptible to the challenges of the protection of critical information infrastructure and reconcile it with the principles on procedural legitimacy. As a conseqeunce EP3R is perhaps less exclusive and attractive for private actors but may prove more effective in the medium term because it holds ICT infrastructure operators accountable to the public good NIS.

Acknowledgements I would like to thank Jean-Pierre Chamoux for his valuable comments on an earlier draft which was presented at the 27th European Communications Policy Research Conference (EuroCPR), 25–27 March 2012, Ghent, Belgium. I am alone responsible for any inaccuracies and interpretations.

References

Literature

Alderson, D., & Soo Hoo, K. (2004). The role of economic incentives in securing cyberspace. Stanford University's Center for International Security and Cooperation.

Andersson, J. J., & Malm, A. (2007). Public-private partnerships and the challenge of critical infrastructure protection. In I. Abele-Wigert & M. Dunn (Eds.), *International CIIP handbook 2006* (Vol. 2). Center for Security Studies, ETH Zurich.

Assaf, D. (2008). Models of critical information infrastructure protection. *International Journal of Critical Infrastructure Protection, 1*, 6–14.

Assaf, D. (2009). Conceptualizing the use of public-private partnerships as a regulatory arrangement in critical information infrastructure protection. In A. O. Peters, L. Koechlin, T. Förster, & G. F. Zinker-nagel (Eds.), *Non-state actors as standard setters*. Cambridge: Cambridge University Press.

Bauer, J. M., & Van Eeten, M. (2009). Cybersecurity: stakeholder incentives, externalities, and policy options. *Telecommunications Policy, 33*(10–11), 706–719.

BBC (2008). Severed cables disrupt Internet. 31 January 2008. Available at http://news.bbc.co.uk/2/hi/technology/7218008.stm. Accessed 14 February 2012.

Brunner, E. M., & Suter, M. (2008). *International CIIP handbook 2008/2009*. Center for Security Studies (CSS), EHT Zurich.

EP3R (2010). Non-paper on the establishment of a European public-private partnership for resilience (EP3R). Version 2.0, 23 June 2010. Available at http://ec.europa.eu/information_society/policy/nis/docs/ep3r_workshops/3rd_june2010/2010_06_23_ep3r_nonpaper_v_2_0_final.pdf. Accessed 14 December 2011.

Dunn-Cavelty, M., & Suter, M. (2009). Public-private partnerships are no silver bullet: an expanded governance model for critical infrastructure protection. *International Journal of Critical Infrastructure Protection Issue, 4*(2), 179–187.

Fieldes, J. (2011). Stuxnet virus targets and spread revealed. 15 February 2011. Available at http://www.bbc.co.uk/news/technology-12465688. Accessed 14 February 2012.

Hämmerli, B., & Renda, A. (2010). *Protecting critical infrastructure in the EU*. Regulatory policy. CEPS Task Force Reports, Centre for European Policy Studies (CEPS).

Héritier, A., & Eckert, S. (2008). New modes of governance in the shadow of hierarchy: self-regulation by industry in Europe. *Journal of Public Policy, 28*, 113–138.

Hill, M. (2005). *The policy process* (4th ed.). London: Longman.

Hood, C. (1991). A public management for all seasons? *Public Administration, 69*(1), 3–19.

ITU (2011). *ITU national cybersecurity strategy guide*. Geneva: ITU. September 2011. Available at http://www.itu.int/ITU-D/cyb/cybersecurity/docs/ITUNationalCybersecurityStrategyGuide.pdf. Accessed 11 December 2011.

Lane, J.-E. (2009). *State management*. London: Routledge.

Maurer, T. (2011). *Cyber norm emergence at the United Nations—an analysis of the UN's activities regarding cyber-security?* Discussion Paper 2011-11, Cambridge, Mass.: Belfer Center for Science and International Affairs, Harvard Kennedy School, September 2011.

Menn, J. (2010). Police shut down Mariposa hacker ring. Financial Times. 3 March 2010. Available at http://www.ft.com/intl/cms/s/0/f6960e5a-2711-11df-b84e-00144feabdc0.html#axzz1otZbvVuh. Accessed 14 February 2012.

Moore, T. (2010). The economics of cybersecurity: principles and policy options. *International Journal of Critical Infrastructure Protection, 3*(3–4), 103–117.

OECD (2001). The DAC Guidelines, Poverty Reduction, OECD, Paris. Available at http://www.oecd.org/dataoecd/47/14/2672735.pdf. Accessed 11 December 2011.

OECD (2002). OECD guidelines for the security of information systems and networks: towards a culture of security. Available at http://www.oecd.org/dataoecd/16/22/15582260.pdf. Accessed at 14 February 2012.

OECD (2003). Policy coherence: vital for global development. Policy Brief, OECD, Paris. Available at http://www.oecd.org/dataoecd/1/50/8879954.pdf. Accessed 11 December 2011.

OECD (2006). The development of policies for the protection of critical information infrastructures (CII): DSTI/ICCP/REG(2006)15/FINAL.OECD, Paris.

OECD (2008). The Seoul declaration for the future of the Internet economy. Available at www.oecd.org/dataoecd/49/28/40839436.pdf. Accessed 14 December 2011.

Pierre, J., & Pieters, G. B. (2000). *Governance, politics and the state*. Hunts: Palgrave Macmillan.

Rhodes, R. A. W. (2000). Governance and public administration. In J. Pierre (Ed.), *Debating governance: authority, steering, and democracy* (pp. 54–90). Oxford: Oxford University Press.

Sabatier, P. & Jenkins-Smith, H. (Eds.) (1993). *Policy change and learning: an advocacy coalition approach*. Boulder: Westview Press.

Servida, A. (2010). Towards a modernised network and information security policy for the European Union. Presentation held at the cybersecurity workshop at Central European University and funded by the European Science Foundation, Budapest, 6/7 June 2010. (On file with the author).

Scott, M., et al. (2001). The role of ENISA in contributing to a coherent and enhanced structure of network and information security in the EU and internationally. Study prepared for the European Parliament. Available at http://www.europarl.europa.eu/activities/committees/studies/download.do?language=en&file=42251. Accessed 14 December 2012.

Shore, M., Du, Y., & Zeadally, S. (2011). A public-private partnership model for national cybersecurity. *Policy & Internet, 3*(2), 8. Available at: http://www.psocommons.org/policyandinternet/vol3/iss2/art8. Accessed 14 December 2011.

Official Documents

Council of the European Union (2007). Council Resolution of 22 March 2007 on a strategy for a secure information society in Europe (2007/C 68/01). Official Journal of the European Union C 68/1 f 24.3.2007.

Council of the European Union (2008). Council Directive 2008/114/EC of 8 December 2008 on the identification and designation of European critical infrastructures and the assessment of the need to improve their protection. Official Journal of the European Union L 345/75 of 23.12.2008.

Council of the European Union (2009). Council Resolution of 18 December 2009 on a collaborative European approach to network and information security (2009/C 321/01).

Council of the European Union (2011). Council conclusions on critical information infrastructure protection: achievements and next steps: towards global cyber-security. 3093rd transport telecommunications and energy council meeting Brussels, 27 May 2011.

ENISA (2009). Analysis of member states' policies and recommendation. ENISA.

ENISA (2010). Incentives and barriers to information sharing in the context of network and information security. ENISA.

ENISA (2011). Economics of security: facing the challenges. A multidisciplinary assessment. ENISA.

European Commission (1994). Green paper on the liberalisation of telecommunications infrastructure and cable television networks. Part one, principle and timetable. COM (94) 440 final. 25 October 1994. http://aei.pitt.edu/1093/1/telecom_cable_gp_part_1_COM_94_440.pdf.

European Commission (2001). Communication from the Commission to the European Parliament, the Council, the European Economic and Social Committee and the Committee of the regions on network and information security: proposal for A European policy approach, COM (2001) 298, Brussels, 6.6.2001.

European Commission (2004). Communication from the Commission to the Council and the European Parliament on critical infrastructure protection in the fight against terrorism, COM (2004) 702 final, Brussels, 20.10.2004.

European Commission (2005). Green paper on a European Programme for critical infrastructure protection, COM (2005) 576 final, Brussels, 17.11.2005.

European Commission (2006a). Communication from the Commission to the European Parliament, the Council, the European Economic and Social Committee and the Committee of the regions on a strategy for a secure information society—dialogue, partnership and empowerment, COM (2006) 251.

European Commission (2006b). Communication from the Commission of 12 December 2006 on a European programme for critical infrastructure protection, COM (2006) 786 final, Official Journal C 126 of 7.6.2007.

European Commission (2009a). Communication from the Commission to the European Parliament, the Council, the European Economic and Social Committee and the Committee of the regions on critical information infrastructure protection—protecting Europe from large scale cyber-attacks and disruptions: enhancing preparedness, security and resilience, SEC (2009) 399, SEC (2009) 400.

European Commission (2009b). Commission staff working document accompanying document to the communication from the commission to the European Parliament, the Council, the European Economic and Social Committee and the Committee of the regions on critical information infrastructure protection—protecting Europe from large scale cyber-attacks and disruptions: enhancing preparedness, security and resilience. Impact Assessment (Part 1), COM (2009) 149, SEC (2009) 400.

European Commission (2010). Communication from the Commission to the European Parliament, the Council, the European Economic and Social Committee and the Committee of the regions. A digital agenda for Europe, Brussels, 26.8.2010, COM (2010) 245 final/2.

European Commission (2011). Communication from the Commission to the European Parliament, the Council, the European Economic and Social Committee and the Committee of the regions on critical information infrastructure protection. Achievements and next steps: towards global cyber-security. 31 March 2011, COM (2011) 163 final.

European Forum for Member States (2011). European principles and guidelines for Internet resilience and stability (Version of March 2011). http://ec.europa.eu/information_society/policy/nis/docs/principles_ciip/guidelines_internet_fin.pdf.

European Parliament and the Council (2002a). Directive 2002/21/EC of the European Parliament and of the Council of 7 March 2002 on a common regulatory framework for electronic communications networks and services (Framework Directive) as amended by Directive 2009/140/EC and Regulation 544/2009.

European Parliament and the Council (2002b). Directive 2002/22/EC of the European Parliament and of the Council of 7 March 2002 on the on universal service and users' rights relating to electronic communications networks and services (Universal Service Directive) as amended by Directive 2009/136/EC.

European Parliament and the Council (2002c). Directive 2002/58/ of the European Parliament and of the Council of 12 July 2002 concerning the processing of personal data and the protection of privacy in the electronic communications sector (Directive on Privacy and Electronic Communications) as amended by Directive 2006/24/EC and Directive 2009/136/EC.

European Parliament and the Council (2004). Regulation (EC) No. 460/2004 of the European Parliament and of the Council of 10 March 2004 establishing the European Network and Information Security Agency. Official Journal L 77, 13/03/2004 P. 1–11.

European Parliament, the Council and the Commission of the European Union (2003), Inter-institutional agreement on better law-making, adopted on 16 December, OJ 2003, C 321/01.

High-level Group on the Information Society (1994). Report on Europe and the global information society: recommendations of the high-level group on the information society to the Corfu European Council. Bulletin of the European Union, Supplement No. 2/94 (commonly called "the Bangemann Report"). Available at http://aei.pitt.edu/1199/1/info_society_bangeman_report.pdf. Accessed 11 December 2011.

Chapter 5
Data Insecurity: Scams, Blags & Scalawags

Sylvia Kierkegaard

Abstract Against a backdrop of rising data breaches, this article examines the legal developments in the USA and the European Union regarding breach notification. Both the US and the EU have enacted security breach laws requiring disclosure to consumers when their personal information has been breached. But the legislation clearly needs sufficient teeth such as higher penalties for organizations who sweep data breaches under the rug, monetary compensation to victims, and imprisonment for those who intentionally break data protection laws and enforcement of sanctions. A harmonized data breach notification law for all sectors may still be some way from becoming a reality.

5.1 Introduction

Public confidence has fallen with report of data breaches grabbing headlines of news outlets. The number of data breaches has surged dramatically in recent years causing wide discontent about the way companies have kept data owner's data insecure. Consumers are concerned that they lack control over their personal information, and identity theft has become all too frequent. The fact is that many organizations are not processing data in a fair and proper manner and keeping the details secure.

According to the Leaking Vault (2011), a new data breach study by Digital Forensics Association, data breaches cost organizations $156.7 billion over a six-year period. The study presents data breach information of publicly disclosed data breach incidents collected from 2005 through 2010, including the disclosure of more than 800 million records over that period. The dollar figure did not include the costs that the organizations downstream or upstream incurred, or the losses sustained by the

S. Kierkegaard is editor in chief of International Journal of Private Law.

S. Kierkegaard (✉)
International Association of IT Lawyers, Copenhagen, Denmark
e-mail: sylvia.kierkegaard@iaitl.org

S. Kierkegaard
University of Southampton, Southampton, UK

J. Krüger et al. (eds.), *The Secure Information Society*, DOI 10.1007/978-1-4471-4763-3_5,   117 data breach victims. Further, the report said the data breach cost estimate was low because 35 % of the incidents did not name a figure for records lost.

Data breaches have the potential to negatively affect a corporation's brand equity and reputation. Sony is struggling to redeem its reputation after suffering a string of massive security breaches and badly mishandling the incident. Sony's hacking woes have continued to mount—a million passwords from the Sony Pictures site and 77 million accounts from the PlayStation Network. More recently, Stanford University's hospital has confirmed that the records of 20,000 emergency room patients were available online for almost a year, while nearly five million current and former troops and their family members had their data stolen from a military contractor in September 2011 putting them at risk for identity theft. The data was saved on computer tapes for the US military's TRICARE health system that was stolen from a car and contained medical records of 4.9 million patients at hospitals and military clinics as well as patients' addresses, phone numbers, lab tests, prescriptions and clinical notes (Vijayan 2011). This breach joined a long list of high-wattage episodes in 2011—similar incidents at NASA, PBS and Lockheed Martin.

The endless manifestation of new and high-profile data breaches has prompted the involvement of legislators at state and executive level. A new subset of law developed—data breach notification, that incorporates elements of privacy regulation, consumer protection and corporate governance mechanisms regarding the security of personal information and information systems (Schwartz and Janger 2005). Data breach notification laws typically require covered entities to implement a breach notification policy, and include requirements for incident reporting and handling and external breach notification.

Federal laws regulating data breach and mandating data breach notification are now being debated in the US Congress. In May of 2011, the Obama administration delivered a cyber security proposal to Congress which would require companies to report data breaches based on a national standard, toughen penalties for computer crimes and direct the Homeland Security Department to work with banks, utilities and transportation operators to develop cyber security plans. The EU has recently implemented the data breach notification law and there are plans to introduce mandatory breach notification in all sectors prompting other countries to propose similar measures. In Australia, the Australian Law Reform Commission's (ALRC) mammoth review of Australian privacy law recommended the creation of an Australian data breach notification requirement, to be implemented through the *Privacy Act 1988 (Cth)*. Taiwan's revised Data Protection Act, which is not yet formally effective, is the first privacy-specific statute in the APAC region to contain an enforceable requirement to notify individuals of a data breach incident. To date, no other privacy legislation in the Asia region has imposed an enforceable legislative requirement to communicate a data breach incident to individuals. The relevant provision requires that, where a public or private sector agency "violates any provision" of the Act, "such that personal data is stolen, disclosed, altered or otherwise impaired," then the agency, after investigating shall notify the subjects by appropriate means. However, the requirement does not extend to every breach occurrence.

With identity theft on the rise and heightened interest in the security of sensitive personal information, many consumers and civil society are now demanding their respective governments to introduce data breach notification clauses. This article will discuss and analyse the current legal developments in data breach notification in the EU and the US.

5.2 Origin of the Data Breach Notification Law

The focus on data breach notification can be traced to California's data breach notification law which was enacted in 2002 and went into effect in July 1, 2003. At that time, California was the only US state with notification laws.

In 2005, Choice point, one of the largest data aggregators and reseller in the US, suffered breach security when 160,000 records were impacted by identity thieves who established bogus accounts. It compiles, stores, and sells information about virtually every US adult. The company disclosed the breach as required by California's Notice of Security Breach law. As a result of the high profile breach and in the absence of a comprehensive federal data breach notification law, the majority of states have passed bills or introduced legislation based on the basic tenets of California's ground breaking original law to require businesses and/or government agencies to notify persons affected by breaches involving their sensitive personal information, and in some cases to implement information security programs to protect the security, confidentiality, and integrity of data.

California's notification requirements apply to any person, state agency, or business that owns, licenses, or maintains computerized data that contains the unencrypted personal information of a California resident.

Civil Code Sec. 1798.82 states:

1798.82. (a) Any person or business that conducts business in California, and that owns or licenses computerized data that includes personal information, shall disclose any breach of the security of the system following discovery or notification of the breach in the security of the data to any resident of California whose *unencrypted* personal information was, *or is reasonably believed* to have been, acquired by an unauthorized person. The disclosure shall be made in the most expedient time possible and without unreasonable delay, consistent with the legitimate needs of law enforcement, as provided in subdivision (c), or any measures necessary to determine the scope of the breach and restore the reasonable integrity of the data system.

(b) Any person or business that maintains computerized data that includes personal information that the person or business does not own shall notify the owner or licensee of the information of any breach of the security of the data immediately following discovery, if the personal information was, or is reasonably believed to have been, acquired by an unauthorized person.

California has since broadened its law to include compromised medical and health insurance information. Although this law requires that notices be sent to affected persons in the most effective and expeditious manner as is reasonable, it says very little about the required contents of such notices. On August 31, 2011, California amended its existing security breach notification law when Governor Jerry Brown signed into law Senate Bill 24 ("SB 24"). The new law, SB 24, updates California's

2002 data breach notification law, which did not contain rules about what information should be include in notification letters. SB 24 establishes rules for what information must be in the data breach notification letter, including a general description of the data breach incident, the type of information breached, the time of the breach, and contact information for major credit reporting agencies. In addition, the new law requires organizations to send an electronic copy of the data breach notification letter to the state attorney general, if a single data breach affects more than 500 Californians. In California, there is a private right of action, and there are very few exemptions.

5.2.1 US Laws

To date the District of Columbia, Puerto Rico the Virgin Islands and 46 states have a data breach notification laws with Alabama, Kentucky, New Mexico and South Dakota holding out (see Fig. 5.1). In 2011, at least 14 states introduced legislation expanding the scope of laws, setting additional requirements related to notification, or changing penalties for those responsible for breaches (NCSL 2011). The recent Amendment to Texas breach notification law (H.B. 300) now extends the breach notification obligation to 50 states. It specifically requires notification of data breaches to residents of states that have not enacted their own data breach notification law, that is, Alabama, Kentucky, New Mexico and South Dakota.

Notification laws vary significantly from state to state. While many US states adopt very similar provisions to the Californian law, some set a different test of when notification will be required (see Fig. 5.1). The notification trigger is the statutory requirement that indicates when and in what circumstance notification is required from an organization. The data breach notification laws in each state define the type of personal information that, when leaked, may give rise to the obligation to notify.

California has a low triggering threshold. Notification of the breach has to be given to any resident of California whose unencrypted personal information was, *or is reasonably believed* to have been, acquired by an unauthorized person—that is, no actual evidence is needed. California, do not require notification where the personal information that was the subject of the unauthorized acquisition was encrypted.

Other states set a different standard as notification is only required in situations where a risk assessment determines that a risk of harm exists to consumers In general, most breach notification statutes do not require notification if there is not a reasonable likelihood that harm to the affected consumers will result. Specific "risk of harm" criteria vary greatly among states. In some states (e.g. Oklahoma, Ohio etc.) harm is limited to identity theft or fraud, while in Washington, the triggering event includes all electronic data that compromises the security, integrity and confidentiality of personal information. Tennessee requires that the unauthorized acquisition of computerized data materially compromises security, confidentiality or integrity of personal data.

All state breach notification laws except for Wyoming's do not require notification if the information is encrypted. Some states (e.g. Rhode Islands, Utah, Vermont

No Law yet enacted

Law enacted

Fig. 5.1 46 States with Breach Notification Law as of 2011

etc.) make exceptions to notification requirements if law enforcement officials determine it will interfere with a criminal investigation.

Many states provide a safe harbour for an entity that is regulated by state or federal law and maintains procedures pursuant to such laws, rules, regulations, or guidelines. Reportedly 29 states impose similar duties for the public and private sectors, 14 states do not, and Oklahoma's law applies only to the public sector (Stevens 2010).

There are also differences in terms of deadlines for notification, definitions and civil penalties.

The general approach adopted in a number of states, including California, is to define personal information as an individual's first name (or initial) and last name, in combination with any of the following: social security number; driver's licence number or state identification card number; or account number, credit card number or debit card number in combination with any necessary security code, access code or password that would permit access to the account as well as medical and health information. Some US states include biometric data in the definition of 'personal information.' Wisconsin and North Carolina, for example, includes DNA profile, fingerprint and biometric data while Delaware's definition of personal information includes 'individually identifiable information, in electronic or physical form, regarding the Delaware resident's medical history or medical treatment or diagnosis by a health care professional (Data Quality Campaign 2011).

Although most states require notification of affected individuals in the most expedient manner possible, three states (Florida, Ohio and Wisconsin) require notification within 45 days of discovery.

In many states, non-compliance with breach notification law carries civil or criminal penalties of varying types and degrees. In Virginia, there is no civil penalty unless the court finds that the defendant has engaged in a course of repeated and wilful violations. Civil penalty cannot exceed $150,000 per breach in contrast to Washington which allows a civil penalty of not more than $100 for each violation.

Most breach notification laws apply only to electronic records; fewer than 10 states specifically contemplate notification for breaches of paper records. In many states, noncompliance with breach notification law is offered a private right of action. In all US states, the responsibility for deciding whether notification is required following a breach in the security of the system rests with the organization itself.

This confusing patchwork of distinct standards has highly uneven results and many gaps in coverage. Since the data breach requirements vary by state, this slows companies down when trying to determine the correct post-breach course of action.

5.2.2 Legislative Proposals

The increasingly complex and diverse state data breach notification laws have prompted both the House and Senate to propose a new comprehensive federal data breach notification law that would pre-empt stronger state laws. The Senate Judiciary Committee has approved along partisan lines 3 new bills that deal with data breach. The three bills, proposed by Chairman Leahy (D-VT), Senator Blumenthal (D-CT), and Senator Feinstein (D-CA), would require businesses to develop data privacy and security plans and set a federal standard for notifying individuals of breaches of sensitive personally identifiable information. The Leahy (S.1490) Personal Data Privacy and Security Act of 2009 and Feinstein (S.139) Data Breach Notification Act bills which would apply to business entities engaged in interstate commerce and require data security programs and notification to individuals affected by a security breach. It would relieve businesses and agencies from breach notification if they conduct a risk assessment and conclude there is no significant risk of identity theft, economic loss or physical harm to individuals by the breach; the Blumenthal bill has the same formulation but refers simply to harm generally. Under all three bills, if businesses and agencies conclude there is no significant risk of harm arising from the breach, they must share the results of the risk assessment with the Federal Trade Commission (Geiger 2011).

In the lower house, the House subcommittee has approved the SAFE Data Act (HR2577). The breach notification provisions of the Act require companies to notify law enforcement without unreasonable delay and notify the FTC and all affected individuals whose personal information "may have been accessed or acquired" within 48 hours of identifying the affected individuals. The notification to affected individuals must begin no later than 45 days after discovery of the breach unless the company receives a written request to delay notification by law enforcement.

Notice to affected individuals is required when there is unauthorized access to or acquisition of personal information in electronic format. Personal information is limited to a person's name in combination with a: (1) Social Security number; (2) driver's license number, passport number, military ID; or (3) financial account number or credit or debit card number along with any required code necessary to permit access to the account. There is also risk of harm trigger—notice is not required if the company makes a reasonable determination that the breach presents "no reasonable risk of identity theft, fraud, or other unlawful conduct" to the affected individuals. A presumption exists that there is no reasonable risk of harm if the data was encrypted.

The definition of personal information in the bill is "far too limited," and does not does not protect personal information such as e-mail addresses, payroll records, and online pictures and videos. The proposal addresses only the risk of identity theft and financial harm and excludes sensitive information beyond those types of data that can be used to perpetuate financial fraud.

5.2.3 Federal Law

Breaches are regulated by states, with the exception of health information breaches.

While organizations are subject to differing data breach notification requirements, depending on their state of operation, all financial institutions and health care providers in the US are subject to the data breach notification requirements.

Federal laws, such as the Health Insurance Portability and Accountability Act (HIPAA) and the Gramm-Leach-Bliley Act, govern some sectors such as the health care industry and financial institutions. They require private sector covered entities to maintain administrative, technical, and physical safeguards to ensure the confidentiality, integrity, and availability of personal information.

The Financial Modernization Act of 1999, also known as the "Gramm-Leach-Bliley Act" or GLBA, protects consumers' non-public personal information maintained by a covered financial institution. All financial institutions in the US are subject to the data breach notification requirements set out in the *Interagency Guidance on Response Programs for Unauthorized Access to Customer Information and Customer Notice*, issued by the US Department of Treasury and other agencies (US Interagency Guidance). The US Interagency Guidance interprets the requirements of the *Gramm-Leach-Bliley Act of 1999* (US). The Guidance requires covered financial institutions to notify any customer whose non-public personal information has been subject to unauthorized access or use if misuse of the customer's information has occurred or is reasonably possible.

The Health Insurance Portability and Accountability Act of 1996 (HIPAA) is intended to protect the privacy and security of protected private health information maintained by most healthcare providers (i.e., those who use HIPAA-mandated electronic codes for billing purposes), health insurance companies, and employers who sponsor self-insured health plans. The two principal sets of regulations issued by HHS to implement HIPAA are the Standards for Privacy of Individually Identifiable Health Information (the "HIPAA Privacy Rule") and the Security Standards for Individually Identifiable Health Information (the "HIPAA Security Rule").

The Health Information Technology for Economic and Clinical Health Act (HITECH Act) became a law on February 17, 2009, and took effect on February 17, 2010. It supplements the requirements of the HIPAA Privacy Rule and the HIPAA Security Rule. Section 13402 of the HITECH Act requires a covered entity to notify affected individuals when it discovers that their unsecured protected health information (PHI) has been, or is reasonably believed to have been, breached.

The notification of a breach must include a description of what occurred; the types of information involved in the breach; steps individuals should take in response to the breach; what the covered entity is doing to investigate, mitigate, and protect against further harm; and contact information to obtain additional information. The Act provides exceptions to this definition to encompass disclosures where the recipient of the information would not reasonably have been able to retain the information, certain unintentional acquisition, access, or use of information by employees or persons acting under the authority of a covered entity or business associate, as well ascertain inadvertent disclosures among persons similarly authorized to access protected health information at a business associate or covered entity. The health plan, health care provider, or business associate will be required to give notice of the breach without unreasonable delay, and no later than 60 calendar days after its discovery.

A "breach" is defined as the "unauthorized acquisition, access, use, or disclosure of protected health information" *which compromises the security or privacy of such information*, except where an unauthorized person to whom such information is disclosed would not reasonably have been able to retain such information (Section 13400(1) of the Act). The definition of "compromises the security or privacy" language contemplates that covered entities will perform some type of risk assessment to determine if there is a risk of harm to the individual, and therefore if a breach has occurred. The Interim Final Rule of the Act agreed that the statutory language encompasses a harm threshold and have clarified in paragraph (1) of the definition that *"compromises the security or privacy of the protected health information"* means *"poses a significant risk of financial, reputational, or other harm to the individual."* Thus, to determine if an impermissible use or disclosure of protected health information constitutes a breach, covered entities and business associates will need to perform a risk assessment to determine if there is a significant risk of harm to the individual as a result of the impermissible use or disclosure.

Privacy incidents involving protected health information (PHI) occur all the time and according to the vaguely specified "harm threshold," may not qualify if they do not pose "significant harm." The problem is that determining the harm can be complex. Relying on an organization to determine risks in a regulatory vacuum, results in inconsistencies. *How are organizations able to evaluate the harm of their own breaches?*

Despite the unprecedented challenges to privacy in the modern environment, there is still no comprehensive law that spells out consumers' privacy rights. Overall, the US approach focuses more closely on industry specific uses of information like credit reports or medical data rather the on the protecting the privacy of individuals. There must be a way to broaden the definition of private data and consolidate private data security and breach notification legislation to cover all sectors.

5.2.4 Class Action

In what may be considered the first case on data breach admitted in the US Federal Court, the 1st US Circuit Court of Appeals in Boston allowed negligence and contract punitive class action litigation to proceed in a grocery store data breach because of the alleged damages incurred.

In *John Anderson et al. vs. Hannaford Brothers A/S*, a class action suit was filed against Hannaford Brothers, owners of Scarborough Maine grocery chains, after plaintiffs experienced more than 1800 unauthorized charges to their accounts after making purchases at the stores. The breach affected more than 4.2 million credit and debit cardholders in 270 stores.

Until the ruling, class actions relating to data breaches have generally been dismissed by the courts because plaintiffs do not have the locus standi or there was no threat or actual damage (Greenwald 2011).

The court's decision signals the court's willingness to take the issue of data breach more seriously.

5.3 European Data Breach Law

Directive 2009/136/EC amended Directive 2002/58 (Privacy Directive) concerning the processing of personal data and privacy. It introduced an obligation to notify individuals and Authorities in instances of information security breaches.

According to the EU Communication, security breach notifications serve different functions and strive for different goals: to serve as an information tool to make individuals aware of the risks they face when their personal data are compromised so that they can take the necessary preventive measures to mitigate risks (such as changing passwords), encourage data controllers to implement stronger security measures to prevent these breaches, and enhance the enforcement powers of data protection authorities.

The scope of the E-Privacy Directive includes providers of public electronic communication networks and services such as telecom operators, mobile phone communication service providers, Internet access providers, providers of the transmission of digital TV content (not the content providers), and other providers of electronic communication services. The European legislature explicitly stipulates that the directive "does not apply to closed user groups and corporate networks" (Recital 55).

Personal data breach is defined as "*a breach of security leading to the accidental or unlawful destruction, loss, alteration, unauthorized disclosure of, or access to, personal data transmitted, stored or otherwise processed in connection with the provision of publicly available electronic communications service in the Community."*

According to Art. 4(3), the provider shall, *without undue delay*, notify the personal data breach (without any threshold of harm) of any personal data (including protected/encrypted data) to the competent national authority.When the personal data breach is likely to *adversely affect* the personal data or privacy of a subscriber or individual, the provider shall also notify the subscriber or individual of the breach without undue delay. A breach adversely affects the data or privacy of a subscriber or individual where it could result in identity theft, fraud, physical harm, sign humiliation or damage to reputation (Recital 61). *It is not clear if the organization is responsible for making the risk assessment.*

The ePrivacy Directive allows the Commission to propose 'technical implementing measures'—practical rules to complement the existing legislation. The member states are encouraged to adopt measures delineating the circumstances, format and procedures for information and notification requirements, granting them increased powers to control the notification process.

They shall also be able to audit whether providers have complied with their notification obligations under and shall impose appropriate sanctions in the event of a failure to do so.

Notification of a personal data breach to a subscriber or individual concerned shall not be required if the provider has demonstrated to the satisfaction of the competent authority that it has implemented appropriate technological protection measures, and that those measures were applied to the data concerned by the security breach. Such technological protection measures shall render the data unintelligible to any person who is not authorized to access it.

The notice contents differ depending on the notice recipient, i.e. the authorities or the individuals. The notification to the subscriber or individual shall at least describe the nature of the personal data breach and the contact points where more information can be obtained, and shall recommend measures to mitigate the possible adverse effects of the personal data breach. The notification to the competent national authority shall, in addition, describe the consequences of, and the measures proposed or taken by the provider to address, the personal data breach.

Providers shall maintain an inventory of personal data breaches comprising the facts surrounding the breach, its effects and the remedial action taken which shall be sufficient to enable the competent national authorities to verify compliance.

In order to facilitate better enforcement and compliance with the E-Privacy Directive, the Authorities are granted increased enforcement powers with respect to the security of processing.

5.3.1 Varying Implementations

A growing number of Member States have now implemented the data breach notification law. The implementation, however, vary widely from member state to member state. Some members have broadened the scope of the directive to include all sectors (Germany), while others (Hungary) limit the notification obligation to electronic communication provider.

In Ireland, providers of publicly available electronic communications must notify the Office of the Data Protection Commissioner of the breach (*even in circumstances where it considers the data would be unintelligible to third parties*) including a description of the measures to be taken to address the breach and notify any individual that may be adversely affected by the breach. (Privacy and Electronic Communications) Regulations 2011 (SI 336 of 2011). In case of any particular risk to the security of the network, providers of publicly available electronic communications networks or services must provide information to subscribers *without delay* (*within two working days of becoming aware of the incident*) about the risks and any possible remedies (including the likely costs involved) even where the proposed measures are outside the direct control of the undertaking. It is not necessary to notify individuals if the Office of the Data Protection Commissioner is satisfied that the data would be unintelligible to third parties.

On 10 June 2011 Hungarian Parliament received the draft bill on the amendment of the Act No. C. of 2003 on electronic communications, which transposes the personal data breach provisions of the recently amended European ePrivacy Directive. Once the service provider reports a data breach and it did not provide a notice on data breach to its customers, the Hungarian Media and Communications Agency may—upon consideration of the risk of possible detrimental consequences of the security breach—imposes an obligation on the service provider to notify its customers after the Agency has obtained the opinion of the Data Protection Commissioner. The Agency would be empowered to release guidance on data breach notification as well as on the best practices relating to the security requirements of data processing. However, the new Hungarian Data Protection Act (Act No. CXII of 2011 on Informational Self-Determination and Freedom of Information) which will enter into force in January 2012 has no data breach notification provision.

On 1 January 2010, Austria introduced a "data breach notification duty" in the 2010 Amendment to the Data Protection of Act. Paragraph 2a of Sect. 24, private enterprises and public agencies are obliged to inform data subjects if they become aware of "*systematic and seriously wrongful use of data*" where the data subject may be harmed. However, this duty for information does not take effect if the breach will only cause minor harm (geringfügiger Schaden) (Paragraph 2a of Sect. 24). The limitation to "systematic and serious misuse" suggests that when the breach is incidental, notification is not necessary. Notification is only required when the organization knows about the breach. In addition, notification is not required in the case of any of the following: the data breach only results in non-economic damage; potential damage is minor and the cost of informing all individuals would be disproportionate.

France's new law (Ordonnance n° 2011–1012 du 24 août 2011 relative aux communications électroniques, or the "Ordinance") implements the provisions of the revised EU Directive 2002/58/EC (the "e-Privacy Directive") and imposes an obligation on electronic communication service providers to provide notice in the event of a data security breach to the French Data Protection Authority (the "CNIL"). If the breach is likely to impact subscribers' (or any other individual's) right to the protection of personal data or right to privacy, the service provider also must inform the potentially affected individuals without delay. The service provider is not required to inform affected individuals if the CNIL determines that appropriate protective measures have been implemented to render the data in question inaccessible or indecipherable by unauthorized individuals (Information-age 2011; Hunton Privacy Blog 2011).

The UK will not implement the breach notification law (Out-Law 2011). Instead the Information Commissioner Office (ICO) has issued guidelines to businesses when they need to report data breaches to ICO. Government agencies are already required to report data breaches and since 2010, the UK Information Commissioner's Office has had the power to fine all organizations up to £500,000 for data breaches. The size of the imposed fine is proportional to the seriousness of the breach, the organization's financial resources and the sector it serves.

Section 42a of the German Federal Data Protection Act (*Bundesdatenschutzgesetz*) (BDSG) contains a statutory data breach notification requirement to all private organizations (and public entities that compete in the free market). Notification is required for breaches that may lead to "serious impediments for privacy and other individual interests." They must notify without undue delay both the competent DPA and affected individuals of any unlawful transfer or other disclosure of certain types of personal data to third parties under certain circumstances. Relevant circumstantial requirements include the type(s) of data involved and whether there is a threat of serious effects on the rights or protected interests of the data subjects resulting from the transfer or disclosure. The notification obligation is triggered when the breach involves: sensitive data; criminal records; bank account or credit card data; personal data that is subject to legal privilege (for example, data held by lawyers, doctors or journalists); or data collected on users of online services. In cases where a large number of individuals are affected, public announcements in at least two national newspapers may replace individual notices.

Countries also impose varying penalties.

In Ireland, failure to comply with these obligations can result in a criminal prosecution with fines of up to €5,000 and on indictment €250,000 per offence. Noncompliance with these provisions is punishable in France by up to five years of imprisonment and a €300,000 fine. The UK Information Commissioner's Office has the power to fine all organizations up to £500,000 for data breaches in a way that is likely to "cause substantial damage or substantial distress." Despite the ICO being able to issue penalties of up to £500,000, none of the fines issued have been more than £100,000, with the total penalties issued to date standing at £310,000.

German DPAs may impose a fine of up to €300,000 for failure to provide notification of a data breach, or for failing to provide notification correctly, completely, or in a timely manner. The Danish Data Protection Act does contain a penal provision in case of breach of Sect. 5 and 41.3 of the Act. The penalty ranges from a fine up to 4 months of imprisonment. The data subject might bring an action for damages to the courts if certain criteria are fulfilled. However, most of the time they only get a warning, community service or a stay in the rest house. In Estonia, violations of data protection requirements are treated as misdcmcanours and are punishable by a fine. Damages can be sought by individuals under the Czech Republic Civil Code if the breach resulted in the violation of their personality rights. He is entitled to bring civil action. In Spain, fines may be imposed if it is shown that the breach is a result of a breach of the regulations concerning mandatory security measures.

5.3.2 Key Concerns

The picture that emerges is that member states have varied approaches to enforcing breach notification protection while others have not extended the scope of application beyond providers of electronic communication services. Some of the EU member states' texts include various degrees of thresholds to inform individuals.

The Privacy directive grants the authorities more enforcement powers, specifically 'the threshold' that may trigger when to notify. Some might define situations which are below the thresholds, while others will have stricter rules as illustrated in the previous discussions. Notification of every breach has been assailed by numerous quarters for creating extra burden to the companies and "notification fatigue to authorities." Organizations are demanding that EU member states must agree on what constitutes the minimum thresholds for the type of breach, and the numbers affected. Breaches must be categorized according to risk levels to avoid 'notification fatigue'. For example, this could include breaches of personal information, which due to its sensitivity, should be deemed to meet the threshold.

The Article 29 Working Party noted, "Differences may also arise as far as the implementation of the exception relating to technological protection measures, which must render the data unintelligible to any person who is not authorized to access it. Such possible divergences may arise because under Article 4(3) it is for national competent authorities to assess whether the technological measures are appropriate and if they were applied" (Working Document 01/11 2011). Since the ePrivacy Directive provides an exemption from the obligation to notify when data is rendered unintelligible to any person not authorized to access it, different levels of technological protection has emerged as precise measures have not been specified in the Directive.

The language of the Directive is also unclear on what constitutes the "Reporting Delay." A concrete deadline for reporting the breach is needed. Organizations should have less discretion in deciding when and if disclosure must occur. Data subjects want to be notified immediately when a breach occurs in order for them to take concrete steps to mitigate the harm, such as in situations where credit card details have been compromised. Organizations, want a longer deadline, to enable them to make a public relation damage control or to identity-solve the problem. However, this should not hinder them from notifying the data subject immediately.

Questions still remain as to which authority will issue such guideline—the data protection authorities, the e-communications authority or both? Many are complaining that they do not have the experience and resources to deal with breach notifications.

The Directive gives Member States the right to lay down the rules on penalties, including criminal sanctions where appropriate. However, penalties in the form of fines are rarely imposed unless they concern gross breaches of data protection. Google was given only a "slap on the wrist" after its Street View cars collected emails and other data from Wi-Fi networks while driving around UK cities during 2008 and 2009. The company did receive punishment or fine because "it did not cause serious harm."

Although the UK privacy watch dog, ICO, are able to issue penalties of up to £500,000, none of the fines issued have been more than £100,000, with the total penalties issued to date standing at £310,000. Moreover, it has fined less than one per cent of organizations reported for breaching the Data Protection Act (DPA) since it gained new powers in 2010. The Information Commissioner's Office (ICO) issued fines in relation to just four of the 2,565 suspected data breaches reported to the watchdog between 6 April 2010 and 22 March 2011 (Heath 2011) Cosmetics Retailer Lush was not fined even though it was found to have failed in conducting regular security checks thus compromising the details of 5000 customers. The company also failed to process credit card data in accordance with the Payment Card Industry Data Security Standard. The hacking occurred between October 2010–January 11, 2011 and it took the company sometime before notifying the authorities.

Additionally, data subjects whose personal data have been loss, deleted or access without authorization are rarely provided compensation.

Criminal sanctions should also be imposed on companies who deliberately bury the evidence and delay notification. For example, the gambling onsite Betfair reported the security breach involving 2.3 million payments cards only after 18 months after the incident (Kitten 2011). Betfair says it kept the breach under wraps because it had determined internally that no customer data had been harmed. Law enforcement may have exacerbated the problem by recommending that Betfair do not disclose the breach, fearing public knowledge might jeopardize the investigation.

Fines may provide an inadequate deterrent, when the financial rewards for illegal behaviour are so great. Fines will not deter future breaches in the private sector. For example, the £60,000 fine issued to A4e by the UK privacy watchdog following the theft of an unencrypted company-owned laptop from the home of an employee was only a fraction of the company's £145m turnover (Condon 2011).

Imprisonment is needed for people who commit blagging—or selling it on without permission of personal data. In Denmark, some online companies harvest the details of customers and send their information automatically to a debt claim company, who contact the inactive customers after many years with a claim for unpaid debt. The claim company works in partnership with a law firm, who does not need authorization from the police to make inkasso claim. Many people without debt find themselves harass by the law firm.

The misuse of privileged access to personal information is widespread that health personnel's and even police officers access other people's personal information.

A police was fined £1,000 for accessing the personal data of people who had reported crimes. Hullandeastriding (2011), while a nurse who provided patient details
to her partner who worked for an accident management company was only fined
150 per offence, even though such companies pay up to 900 for a client's details
(Nuyen 2011).

The UK National Health System (NHS) has one of the highest rates of breaking UK data laws, with regular data breaches reported by the Information Commissioner's Office and millions of patients records lost or mishandled during 2011
alone. Patients have not been compensated and very few hospitals received fine. According to Freedom of Information Act requests by Big Brother Watch, there were
806 incidents over the last three years where the laws protecting the privacy of patient records were breached. Breaches included 23 instances of patient information
being posted on a social network, 91 incidents of staff looking up colleagues' details, while 24 NHS trusts saw confidential information stolen, lost or left behind
by staff. Of the 800 incidents discovered, just 102 cases resulted in staff dismissal
(Bigbrotherwatch 2011). The leaks are just the tip of the iceberg as many patient
data breaches are alleged to be covered up by the health providers.

Fines are not appropriate in circumstances where the data endanger or trauma-
tized the data subject.

5.3.3 US and EU Data Breach Notification Compared

Organizations in the US have to deal with differing and inconsistent requirements
of the individual state laws. In general, US data breach notification law applies horizontally to all organizations that process certain types of information, while the
EU breach notification obliges only telecomm and Internet service providers. On
one hand the EU has a harmonized mandatory breach notification law; on the other
hand, diverging approaches have emerged in the areas of the scope of application
of the directive, harm threshold, notification procedure and specific technological
enforcement measures allowing exemptions.

Most states that have data breach notification laws, including California, do not
require notification where the personal information that was the subject of the unauthorized acquisition was encrypted in contrast to the EU where breach covers both
encrypted and unencrypted data. The concept of "personal information" is also
much broader in the EU, compared with the United States.

Breach notification is imposed on electronic communication providers for any
loss, modification or destruction, or unauthorized access in the European Union
while the US limits itself to unauthorized access of personal information.

In the US, data breaches make frequent news. Health commissioners, insurance
commissioners and attorneys general in a growing number of states have been initiating enforcement actions to protect the privacy of consumers in their states. Following the implementation of the HITECH health data breach notification rule, the
HHS Office for Civil Rights has stepped up its levying of fines–most notably on

the University of California ($865,500 in July), Cignet ($4.3 million in February), Massachusetts General Hospital ($1 million in February) and RiteAid ($1 million in July 2010). A handful of EU member states also imposed significant fines over the past year. According to the 2011 IAPP benchmarking survey of data protection authorities to be released at the 33rd Annual International Conference of Data Protection and Privacy Commissioners in Mexico City, European DPAs collected more than $31 million in fees in the past year. Just three member states—Spain, Italy and the UK—accounted for nearly this entire amount, however (Cline 2011).

Ultimately while data breach notification laws may have succeeded at highlighting significant problems regarding organization information security failings, they have not really provided effective remedies to resolve those problems—due essentially to their conflicting conceptual base and resultant differences in application (Burdon 2011).

The US approach is sectoral and aims to address ineffective corporation security, while the EU law is privacy-focused. However, while the EU provides a broad privacy rights regarding the collection, storage and use of personal information, it gives only limited rights-based protections to individuals regarding unauthorized disclosures. In spite of high profile case, the lack of effective enforcement, penalties and victim compensation have made the notification law toothless.

Private lawsuits related to data breach incidents are not successful. Many states lack laws permitting private lawsuits for damages related to data loss and where lawsuits are allowed, consumers have difficulty to prove that the breach caused them legally cognizable harm.

Sony fell victim to what has been called the largest data breach ever, affecting nearly 77 million users of Sony's Playstation and Qriocity services by an organized group of hackers. Several putative class actions have been filed against Sony with complaints alleging negligence, invasion of privacy, misappropriation of confidential financial information, breach of implied contract, and breach of express contract. Disgruntled consumers have filed a total of 55 breach-related suits against Sony in the United States alone.

The recent decision of the First Circuit's ruling in *Anderson v. Hannaford Bros. Co.*, Nos. 10-2384, 10-2450 (1st Cir. Oct. 20, 2011) could pave the way for game-changing victory in cases where claims involve actual or legitimate threats of misuse. The First Circuit upheld implied contract and negligence as proper theories of recovery. In regards to damages, the First Circuit reversed the trial court and found that "a plaintiff may recover for costs and harms incurred during a reasonable effort to mitigate." To recover, however, the plaintiffs needed to establish an actual injury such as money lost as opposed to only time and effort. In finding that the plaintiffs stated a proper claim for damages in a data breach case, the First Circuit noted that the Hannaford breach was not inadvertent loss or simple breach with no misuse. Rather, the court emphasized that there was actual misuse of the information that may have been global in reach running up thousands of charges. This type of breach presented a "real risk of misuse."

5.4 Conclusion

The relentless breaches show no signs of slowing down.

The European Commission's proposal for a mandatory breach obligation for all personal data controllers is a step forward. The EU intends to adopt a horizontal breach notification obligation as part of the revision of the Data Protection Directive.

However, the laws are toothless if the organizations are allowed to make their own risk assessment of what breach constitutes serious harms. With companies, criminals and negligent employees merely receiving a gentle slap on the wrist for major data breaches, the real losers are the victims whose personal information have been compromised causing moral, financial and psychological harm without any luck for monetary compensation. This is a trajectory of injustice.

It's about time we put a criminal and their sorry carcasses to jail. Until now, data breach offenders have only meted fines. The wheels of justice grind slowly, but must they grind this slow?

References

Bigbrotherwatch (2011). NHS confidentiality breaches 5× a week. Available at http://www.bigbrotherwatch.org.uk/home/2011/10/nhs-data-protection.html.

Burdon, M. (2011). Contextualizing the tensions and weaknesses of data breach notification and information privacy law, 27. Santa Clara Computer and High Technology L J 63.

Cline, J. (2011). New wave of privacy regulation and enforcement. IAPP.

Condon, R. (2011). First data protection fines issued after UK data breaches. Searc.co. Available at http://searchsecurity.techtarget.co.uk/news/1524282/First-Data-Protection-Act-fines-issued-following-UK-data-breaches.

Data Quality Campaign (2011). State security breach response law. Available at http://dataqualitycampaign.org/files/State%20Security%20Breach%20Chart%20Final%20for%20posting%202011%2003%2010.pdf.

Greenwald, J. (2011). Data breach ruling may signal change in the court's approach. Business insurance. Retrieved 31st October 2011 from http://www.businessinsurance.com/article/20111030/NEWS07/310309999?tags=%7C299%7C256%7C75%7C303%7C335.

Geiger, H. (2011). Senate judiciary passes 3 data security bills. Center for democracy and technology. Retrieved 2 November 2011 from http://www.cdt.org/blogs/harley-geiger/239senate-judiciary-committee-passes-three-data-security-bills.

Heath, N. (2011). Most data breaches escape privacy watchdog fines. Silicon. Available at http://www.silicon.com/technology/security/2011/04/21/most-data-breaches-escape-privacy-watchdog-fines-39747329/.

Hullandeastriding (2011). Police fined £1000 for stealing personal data. Available at http://www.thisishullandeastriding.co.uk/Police-official-fined-stealing-personal-data/story-11978275-detail/story.html.

Hunton Privacy Blog (2011). Available at http://www.huntonprivacyblog.com/2011/08/articles/france-introduces-data-security-breach-notification-requirement-for-electronic-communication-service-providers/.

Information-age (2011). Available at http://www.information-age.com/channels/informationmanagement/news/1650778/france-enacts-breach-notification-law-for-isps-and-telcos.thtm.

Kitten, T. Y. (2011). Online gambling site exposes 2.3 million payment cards. Bank info security. Available at http://www.bankinfosecurity.com/articles.php?art_id=4127.

Leaking Vault (2011). Available at http://www.digitalforensicsassociation.org/storage/The_ Leaking_Vault_2011-Six_Years_of_Data_Breaches.pdf.

NCSL (2011). Security breach legislation 2011. Retrieved 1 November 2011 from http://www. ncsl.org/default.aspx?tabid=22295.

Nuyen, A. (2011). UK MPs call for jail sentences in data breach cases. Computer World.

Out-Law (2011). Available at http://www.out-law.com/page-9619.

Schwartz, P., & Janger, E. (2005). *Notification of data security breaches*, 105 Mich. L.R., (2007); Thomas J. Smedinghoff, *Security breach notification—adapting to the regulatory framework*, 21 Rev. Bank. Financ. Serv. (2005).

Stevens, G. (2010). Federal information security and data breach notification laws. Congressional Research Service.

Vijayan, J. (2011). Data breach affects 4.9M active, retired military personnel. CIO. Available at http://www.cio.com/article/690733/Data_Breach_Affects_4.9M_Active_Retired_Military_ Personnel.

Working Document 01/11 (2011). On the current EU personal data breach framework and recommendations for future policy developments.

Part III
New Technological Cybersecurity

Chapter 6
Content Analysis in the Digital Age: Tools, Functions, and Implications for Security

Johan Eriksson and Giampiero Giacomello

Abstract Content analysis is an established and effective method for research in the social science and, despite what many think, it has been around for quite some time. It has also tremendously benefited from ICT and the growth of computing power, as computers have proved to excel in the dull routine of scanning texts for keywords. But content analysis has become ubiquitous with the advent of the Internet, particularly emails and Web sites. Keyword search, a pivotal element of content analysis, is the most widespread feature of many Internet applications, from search engines to password-cracking programs. Consequently, it has become a central concern for cybersecurity. This chapter investigates some of the most important applications of content analysis on the Net and discusses its increasing essential position in many areas of cybersecurity.

6.1 Introduction

Are people aware of how often they are doing content analysis—that is systematically and quantitatively studying "texts" (Krippendorff 1980)[1]—and do they really understand the ramifications and significance of this method? When we decided that this paper was to be about content analysis, the first thing we did was commencing a content analysis: We simply googled "content analysis". This illustrates our main

---

[1] A texts can be seen as the "material manifestation" of speech. A "step further" from text analysis, the oldest procedure of content analysis, is *discourse* analysis, where the discourse is no longer considered just a simple *reflection* of reality but as its essential constituent part (Phillips and Hardy 2002).

J. Eriksson
Department of Political Science, School of Social Sciences, Södertörn University,
141 89 Huddinge, Sweden
e-mail: johan.eriksson@ui.se

G. Giacomello (✉)
Department of Political and Social Sciences, University of Bologna, Strada Maggiore, 45,
40125 Bologna, Italy
e-mail: giampiero.giacomello@unibo.it

J. Krüger et al. (eds.), *The Secure Information Society*, DOI 10.1007/978-1-4471-4763-3_6,   137
© Springer-Verlag London 2013

point: Content analysis is a ubiquitous activity in the digital age, yet its widespread usage, ramifications and significance have largely gone unnoticed. Content analysis is used not only by intelligence officers and researchers making systematic analyses of texts, but literally by each and everyone using a computer.

The purpose of this paper is to clarify the ubiquity and ramifications of content analysis, the numerous technologies available for content analysis, and the multiple functions of content analysis in the digital age, particularly pertaining to issues of relevance for security. Scientific content analysis using computers and the Internet is growing.[2] Yet we argue that content analysis is something much more fundamental and widespread than the consciously and explicitly scientific content analyses performed by scholars.

In our view, content analysis is an almost inescapable activity in the digital world. With the development of digitalized software in general, and the Internet in particular, huge and constantly growing amounts of text-based information has been made available worldwide. Internet search engines and the inbuilt search functions of browsers and other types of software have become widespread, easily accessible, and user-friendly. Such tools for content analysis are constantly developed and becoming more advanced, while still being very user-friendly. Furthermore, digitalized content analysis does not exclude other methodologies, such as discourse analysis, narrative analysis or argumentation analysis. Such more advanced qualitative methods are, in our view, fully compatible with content analysis. What is unique with content analysis is that it has become so easily accessible, user-friendly, and almost inescapable method in the digital world. It is our contention that—except for more specialized technologies and explicit applications of scientific content analysis—people seldom reflect about that they are using such tools, how they are using them, for what purposes, or with what consequences. Against this background, the paper addresses these more specific questions:

– What tools for content analysis are available in the digital world?
– What are the possible functions and usages of content analysis in the digital world?
– What are the ramifications of such widespread and user-friendly content analysis for security?

The paper is structured as follows. The first section discusses what content analysis is about and how it relates to other methods of textual analysis. The second section provides a brief overview of computer- and web-based technologies for content analysis. The third section discusses the multiple functions, utilizations and types of actors doing content analysis. Finally, in a concluding section, we synthesize our observations and make some recommendations for the development of content analysis, as well as for the study of content analysis and its significance in a digital world. Throughout the paper, we give examples relating to security.

---

[2]For an "early" discussion on this trend, see Weare and Lin (2000).

6.2 The Success of Content Analysis

Content analysis is a widely-employed method for studying political communication and culture (Holsti 1969; Krippendorff 1980; Weber 1990; Druckman 2005).[3] According to Holsti (1969, 25), content analysis can be defined as a "technique for making inferences by systematically and objectively identifying characteristics of specified messages". We may make the mistake of thinking that such technique is a modern "invention", whereas, as Weare and Lin (2000, 272) observe, "outline of content analysis as a systematic and quantitative scientific method for measuring the content of messages have existed for centuries". The rise of newspapers in the 19th century, electronic media in the 20th century and the however, provided methodologists with unprecented amounts of data to be explored via analysis of content. Today, the World Wide Web ubiquity, with its abundance of texts and images, has made content analysis, loosely intended, the principal technique to find information on the Web, whether for legitimate as well as illegitimate purposes. Given these premises, it is not surprising that content analysis is among the principal methods of inquiry recommended and preferred by constructivist scholars in the field of International Relations (Klotz and Lynch 2007).

As Weber (1990, 13) notes, "there is no simple right way to do content analysis". The methods must be defined according to the aims and problems of research. The content analysis can be used for many purposes: to identify the intention of those who communicate a message, to reflect the cultural patterns of groups and institutions, companies or to reveal the focus of certain actors. To start with, a useful differentiation of the unit of analysis is among the sampling, the recording, and the context units (Krippendorff 1980; Weare and Lin 2000). For the former, we focus the *whole* message, for example a newspaper article or a Web page. Recording units are the various, separable *components* in which the message can be subdivided, such as paragraphs, sentences, or single words. In the latter case, we consider the *whole context* in which the message is located, such as, for example, the entire newspaper or Web site.

Next comes the development of categories, such as length, frequency counts, emphasis, qualifications of certain parts and so on (Krippendorff 1980), as well as the training of "coders",[4] who are tasked with dissecting the message according to the categorization scheme. The categories are pivotal for the assessment of validity and reliability of the analysis (Carmines and Zeller 1979; Trochim 1999) and it is crucial to reach the right trade-off between 'reliability' and 'validity' of the categories (Weber 1990; Trochim 1999; Druckman 2005). In the end, the level of internal consistency and accuracy depends on the ambiguity in the words meaning as well as by the choices made by the researcher in defining of the categories and the coding rules.

---

[3]For this section, the authors would like to gladly acknowledge the assistance of Fabrizio Coticchia.

[4]"Human" coders are also fundamental to gauge "each others' (or intercoder) reliability".

As a general example, a common, "three-level" approach to content analysis may include the frequency of the categories, key-words in context and word frequency list. The first level (frequency of categories) analyses how many times in each speech appear the categories of the vocabulary that has been created according to the main frames. Frames are general codes through which actors interpret complex issues. These interpretive schemes simplify external reality through a selective process through which the actors emphasize certain aspects rather than others (Benford and Snow 1992). The elements on which the actors focus on are the most salient in the communication (Entman 1993). Different frames represent alternative ways to address a "theme", which can be almost anything, from, say, "income inequality" to the "rights of humans" to "cyberterrorism" or "life in the universe". Any policy-making process can thus be viewed as a struggle among different frames.

The second level of analysis, which is the KWIC, or key-words in context, illustrates the extracted piece of text where the term is inserted for a length of three lines, allowing a better understanding of its meaning. In fact, a word isolated from their context may cause some mistakes. For instance, identical words may have a different meaning ("leave somebody in peace" vs. "rest in peace") or some terms could be simply negations. Thus, interpretation and selection are essential tasks for guarantying effectiveness of content analysis tools. The third level of analysis is the wordlist: the frequency of words, included in the preliminary vocabulary, used in the text.

In recent years, social scientists have substantially increased the sophistication of content analysis tools, especially thanks to the advent of personal computers and the relevant, specialized software (Weare and Lin 2000; Phillips and Hardy 2002). Software created specifically for content analysis has undoubtedly facilitated the task of the researcher in extremely time-consuming activities such as measuring frequencies and finding bits of information hidden behind considerable amount of "white noise". A frequent criticism towards these methodological tools pinpoints to the lack or underestimation of the "general context" (the context units) in which verbal interactions may occur.

Software programs for content analysis, such as *Hamlet II* or *Words in Context*, allow the researcher to quickly assess the frequency of terms and their location within the text. *Hamlet II*, for example, displays features such as simplicity, immediacy the possession of the necessary requirements for different levels of analysis, which fit well with the aims of most research in the social sciences. Programs such as those mentioned here and others search text files for words or categories in a given vocabulary list, count their joint frequencies within any specified context unit, within sentences, or as collocations within a given span of words and then provide the relevant figures.[5] Words may be classified in the same categories because they may have similar meaning. This 'reduction' and classification process is called 'tagging' (Weber 1990). To operationalize abstract concepts, such as "national interests", "social inequality" or "risk adversion" require to build a category, which is

---

[5]See the Web site of Hamlet II for example, available at http://apb.newmdsx.com/hamlet2.html.

methodologically complex because it requires to capture the essence of the social phenomenon (Druckman 2005).

The use of software can certainly avoid many common methodological problems allowing to apply the coding rules automatically and guarantying accuracy for comparison and reproducibility (Weber 1990). At the same time, computers tend not to be as sensitive to "cognitive differences" as human coders are. As the validity of the content classification obviously depends on the degree of correspondence between categories, words and concepts, it is thus crucial for scholars to be aware of those differences, as well as of the shortcomings of human and computer coding.

Content analysis software has in fact mechanisms that help the researcher to change the vocabulary (e.g. the identification of additional terms) without altering the final outcome. All in all, the construction of a "coding scheme" should always move from two elements: the research question and the texts (content) on which the analysis will be performed. The use of dictionaries of course helps to provide consistency across the categories through synonyms and terms related to the shared theme. Finally, collection and selection of documents are the preliminary tasks before applying the software to the text according to the conceptual categories that have been created through a vocabulary of relevant terms.[6]

6.3 A Web of Tools

Any "keyword-search" can be seen as a sort of content analysis. Admittedly, somebody running a software program trying to guess passwords is not looking for any deeper meaning in the content analyzed or understanding the social implications of the "text" being scanned. Yet, the basic dynamics of a password-cracking application, of a Google search or of a text analysis program processing Dante's *Divina Commedia* or Shakespeare's tragedies are the same, that is, sorting out of the useless "white noise" that important piece of information that the user hoped to find when he or she pressed "enter" in the first place.

Computers, with their specialization in "speeding-up" dull routine tasks have been applied with amazing efficiency to content analysis.[7] Thanks to the Moore's Law, the faster their processors become, the speedier the scanning processes can be. Another general law of computing, namely "garbage in, garbage out", however, fully applies here too. Hence, to find the information we need, we must rely on some analytical thinking in selecting keywords and categories, as the more complex or specialized the information we look for, the deeper and precise should be the preliminary planning and thinking should be. Anyone who simply tried to use

---

[6]We do not make a distinction here between "qualitative" and "quantitative" content analysis, because we argue that such distinction is fictitious as this method is successfully applied in both approaches.

[7]A remarkable example/explanation is given in the reference to "Text Mining" from StatSoft (2011).

the "advanced search" of databases or search engines has experienced this state of affairs.

Computers have then been linked together in networks, multiplying by several degrees of magnitude, their resourcefulness and making them even more indispensable for our societies to function. Upon such networks, a Web of contents, full of images but also of *words*, has been superimposed. Indeed, as the e-mail and the Web are integral and indispensable elements of our lives, we tend to forget that "the Web" was borne as an environment dedicated to hypertext. This is a tool that allows the user to move from one concept to another that is logically "linked" to the former. HTTML is a *hypertext* language, which allows a user to move from one concept to another that is logically "linked" to the former. Web sites are called *content* providers. Or, a "brute force" attacks on trying all of the words of entire vocabularies (in different languages) to guess passwords is just another examples of how, despite all references to the visual experience of the Web, "text" still prevails. Albeit not exhaustive, a list of content analysis-related features would include:

– Internet search engines like Google, Yahoo and older ones like Alta Vista, etc.;
– Search functions in web browsers;
– Search functions on websites;
– Search functions in software applications like Microsoft Office programs, and Adobe Acrobat Reader;
– Specialized software for content analysis like Stata text mining, Hamlet II and others.

Not only do Google and other research engines' algorithms rely on key-words-finding for their search, but the more sophisticated the algorithm, the better the search results, the happier the user, the greater the income from advertising. As the greater part of information provided on the Web is (still) text-based, it is fair to conclude that search engines algorithms are a highly advanced type of content analysis. Because of this state of affairs, software programs capable of screening huge bodies of text (even with different forms) and producing frequency tables and retrieving information have been adopted by the business community as a new aid for its decision-making process and operational research (Berry 2004; Berry and Kogan 2010; Janasik et al. 2009; StatSoft 2011).

Quite simply, this is *text-mining*, the "textual" relation of data-mining (Srivastava and Sahami 2009; Liu 2011). Generally speaking, text mining (sometimes also called intelligent text analysis, text data mining or knowledge-discovery in text) consists of the discovery of previously unknown information from existing resources (Hearst 1999). Unlike data mining, which can detect useful patterns from structured text or data usually stored in large database repositories, text mining searches for patterns in *unstructured natural language* texts (e.g. books, articles, e-mail messages, web pages, etc.). Text mining is generally found useful in environments where large collections of text documents are handled. One of the well-known premises of using text mining is that the value obtained by mining text documents is directly proportional to the value of those documents. The more important the knowledge contained in the document collection, the more value will be derived. Text mining is a multidisciplinary field that includes many tasks such as text analysis, clustering, categorization, summarization, etc.

Technical security reports rely on text analysis and keywords as well (Adevaa and Atxa 2007; Tsai and Chan 2007; Bueno et al. 2011). Last, but not least, as we discuss more in details in the next section, propaganda and perception management, now fully understood by cyberterrorists, are ideal tools to increase the stress conditions of adversary leaders. Such tasks are, more often than not, efficiently carried through by terrorist via images and texts on the Web and social networks (Weimann 2010). Inevitably such items become the primary targets of law enforcement and intelligence services that scan the content in search of useful information.

6.4 The Many Functions of Online Content Analysis

For what purposes can content analysis be used? What the different tools of content analysis have in common is that they are all about selectively and systematically processing text. Beyond this technical level, the functions of content analysis are plentiful. The list below is not comprehensive, but it does include a vast array of different functions which rely heavily on computer-based content analysis:

– Cyber-attacks (e.g. spamming, identity thefts, computer break-ins);
– Intelligence and surveillance;
– Crime investigation;
– Market research;
– Academic research;
– Journalistic investigation;
– Personal purposes (hobbies, travel planning, etc.).

Cyber-attacks such as spamming, denial-of service attacks, identity thefts, and break-ins into computer networks are seldom referred to as forms of content analysis, but that is exactly what they are (Amoroso 2011). E-mail spamming—unsolicited messages repeatedly sent out to massive numbers of recipients—is one of the main forms of Internet disturbance, clogging up e-mail inboxes and often also spreading malicious code. Spamming is preceded by automated online searches for e-mail addresses, or preferably entire e-mail address books. Thus, by asking the search tool to look for text formatted as an e-mail address including the @ sign, culprits are conducting a content analysis. There are several thousands of websites including tools for automated spamming and many other forms of cyber-attacks, and instructions on how to use them (Cordesman 2002, 11).

A common element in e-mail spamming is theft of e-mail addresses which are then used for sending out spam mail. There are many other forms of identity theft, and they are all based on content analysis. After turning a computer on, one of the first things a user has to do, particularly when connecting to a network, is to type in a pass*word*. Hackers can run specific "password-guessing" applications to retrieve the password. Such tools often rely simply on dictionaries (which can include several different languages), trying all the words in those dictionaries, until they possibly find the password. It is for this reason people are recommended to use complicated passwords, combining letters and numbers, and which are not found in any dictionary. Yet there are also "hybrid" attacks which insert not only words but also numbers (0–9) and special characters (such as # or ∞). In all their simplicity methods of password guessing are forms of content analysis: they search for keywords, which can be numerous, and which can contain letters, numbers and special characters. If successful, password guessing could yield illegal access not only to e-mail, but also bank accounts, medical records, military plans, and even operative systems of for example power plants or traffic controls.

Intelligence and surveillance have increasingly become based on ICTs. In a sense, such activities are the other side of the coin of cyber-attacks: in cyberspace, what is a means of defence can easily become a means of attack, and vice versa. Electronic surveillance comes in many shapes and guises. On a global level there are major systems such as the multinational Echelon surveillance system, and the various electronic eavesdropping systems used by for example the US National Security Agency. These systems use satellites as well as online tools for retrieving information from e-mail messages, Internet chat forums, and telephone conversations. Again, these highly advanced technologies apply a very simple tool: keywords. Rest assured that words like "bomb" and "Jihad" are included in these lists of keywords. Likewise, firewalls are supposed to block or filter communication through content analysis. This is achieved through comparing signatures and codes of received packets, or IP addresses of senders, with list of banned content and identities. The functions of firewalls used in personal computers are similar to the technologies used by some national governments to block or filter access to Internet content. It is noteworthy that governments using such censoring measures, especially China, are collaborating closely with private business and media entrepreneurs (Lagerkvist 2011).

Crime investigation is today almost ubiquitously involving ICTs. This includes searching through computer hard disks, e-mail and other forms of electronic communication in which the suspect has been involved (Casey 2011; Zheng et al. 2003). The material to be analysed is often too large to be studied in its entirety. Therefore, automated means of systematically yet selectively retrieving information are used, which typically boils down to the use of keywords.

When it comes to cyber-attacks, however, computer crime investigators and IT security administrators tend to focus on forensics: back-tracing direction of attacks by examining log files and IP addresses (Casey 2011). In such instances content analysis can sometimes be neglected, because they are trying to identify the culprit behind a specific attack, rather than investigating the pattern of activities of culprit and victims. Doing the latter is obviously more easily said than done. It is often impossible to get an understanding of the pattern of activities before the culprit has been identified, or at least identified as a suspect. Yet it can be argued that content analysis is not only a logical step following initial forensics in computer crime investigation, but could also be a parallel means of investigation from the onset. The advantage of this is to get a better contextualization of an individual crime, through a systematic mapping of activities, interests and contacts. This requires close cooperation between crime intelligence units, and units investigating particular incidents.

We suspect that intelligence units (and thus content analysis) tend to play an initial role when automated searches retrieve major hits for particularly sensitive strings of keywords, which points to particular groups or individuals. In other words intelligence analysis, and content analysis in crime investigations more generally, is not only useful for preventing crime, but also for investigating criminal acts after the fact.

Market research is another field in which content analysis is widely used. Market research involves not only analysis of where, how much and by whom particular products and services are purchased. Equally important is analysis of the potential or expected development of markets. This may for example include investigating Internet search patterns. The number of times a particular product or service is "googled", for example, could be an element of market research. Indeed, market research is a growing business area in itself, which is largely and sometimes exclusive Internet-based. Market research companies sell analyses of recent developments, trends and possible futures in particular branches or wider market segments, and can also be commissioned for market analysis for a particular company. There are also freely available market research tools for personal use, such as online price comparison services.

Academic research relies heavily on computerized content analysis. It does so in two ways: on one hand the comparatively limited use of *scientific* tools and methods of content analysis (including software such as Hamlet II, as discussed above), and on the other hand the much more commonplace application of rudimentary content analysis, which is not necessarily guided by any scientific standards. Searching for publications is a major and almost daily activity for researchers, whereby scholars use tools and databases such as Google Scholar, Jstor, the ISI Web of Science, as well as commercial retail sites such as Amazon. Many of these tools, such as Google Scholar, provide not only a list of "hits" for keywords, but also contextual information, such as "similar topics", and where, by whom and how many times an identified publication has been cited.

The development of computerized databases available online has even led to the development of a new academic branch—Bibliometrics—which is completely devoted to a particular form of content analysis, a set of methods to quantitatively analyse scientific and technological literature. Bibliometrics has also become the major standard by which scholars and their work is evaluated. The number of citations in prestigious journals is clearly a dominant criterion, which has also been the target of critique, since citation statistics do not necessarily say anything about the quality or originality of research. It has been argued that the gate keeping powers of and dominant paradigms in scientific journals tend to prevent a lot of critical work to get published (McGinty 1999). This may also say something about the general weakness of computerized and quantitatively oriented content analysis, a topic which however cannot be developed further here.

Journalistic investigation, much like market research and academic research, relies heavily on searches of the World Wide Web, as a significant addition to other journalistic methods (especially interviews). "Googling", we suspect, is an essential activity in contemporary journalism. There is a degree of awareness of this in the media industry as well as among scholars of media and communication, although we suspect that most journalists never reflect seriously on how they are using the Internet, and what the ramifications of their methods are. With the Internet's unbeatable availability, immediacy, global reach, perceived comprehensiveness, and its many user-friendly search engines, there is hardly much incentive to start asking why we are "googling" and with what effects. The dominance of "googling" in journalism has made some critics talk about "robotic journalism" (Kurtz 2002), and "the flattening of expertise" (Brabazon 2006). While "robotic journalism" refers to the automated and unreflective use of Google, "the flattening of expertise" is about how the proliferation of blogs, Twitter, Facebook, and other new online social media challenges the very notions of journalism, journalist and expert. The recent rise of the *individual* as not only a consumer but also provider and distributor of information, arguments and ideas implies the possibility of turning each and everyone into a "journalist", provided they are connected to the Internet.

Finally, computerized content analysis is used for a variety of *personal purposes*. Planning and booking travel, and searching for information about hobbies and contacts within communities of interest imply punching in keywords in more or less advanced search engines. Booking flights, for example, can nowadays be done by using one of the great many online booking agencies. When planning and booking a trip online, the online booking engine typically ask you to submit words for the locations of departure and arrival (and it usually gives you alternatives if you make spelling errors or if there is no immediate "hit"). This is yet another example of how we regularly use online content analysis, but usually without reflecting on its ramifications. What if there are airports, airlines, dates, price levels and other forms of travel not covered by the booking engine you are using? There is a strong tendency, we suspect, to simply ignore such worries, and simply continue using whatever tools we find, have been told to use, or use simply out of habit. This is also of interest from a security perspective, which refers back to the above discussion on intelligence, surveillance and crime: your search patterns can be traced and analyzed, resulting in a "profile". Whether this gives an accurate or interesting picture of whom you are and what you do is an entirely different question.

6.5 Conclusion

Content analysis, which was once a particular, quantitative method of textual analysis among many others, has become a dominant, even ubiquitous way of getting information in the digital age, as keyword search, a pivotal element of content analysis, is the most widespread feature of many Internet applications, from search engines to password-cracking programs. In order to see this and understand its implications, the notion of content analysis must be broadened beyond the limited and demanding concept of scientific content analysis. There are today a great many user-friendly content analysis tools, some merely simple search engines (such as the search tool within MS Word and other software programs), more advanced search engines such as Google, and more specialized tools such as the Hamlet software. Content analysis is also widely applicable, used in many different contexts, for a great variety of purposes. The functions of content analysis includes cyber-attacks such as e-mail spamming, intelligence and surveillance methods, cyber crime investigation, market research, academic research, journalistic investigation, and personal purposes such as the booking of flights.

There are two obvious advantages of computerized, online content analysis: First, it is easily available, user-friendly, yields immediate results, can be applied to vast amounts of material, and has a global reach. Second, computerized content analysis avoids human errors in assembling and coding data. With the exception of technical "bugs", engines for content analysis do what you tell them to do.

That they do exactly what you tell them to do is also the major weakness of computerized content analysis, however. The results of content analysis depend entirely on what keywords are used, and how the results are interpreted. Selecting keywords and interpreting results is not something that content analysis itself can help with. This is where theory, ideological preferences, taste and imagination come in. While this interpretative dimension is understood in the literature on scientific content analysis, we fear that it is far less understood or appreciated among the everyday "googlers" and other users of rudimentary content analysis, including scientists.

If contemporary society so heavily dependent on digitalized and quantitatively oriented content analysis, what does this say generally about we information and knowledge dealt with? It could be argued that while much time and money is spent on developing advanced technologies for content analysis, much less time and money is spent on how we use these tools, what their strengths and weaknesses, and how and with what consequences we choose some keywords before others. In particular, as users of scientific content analysis know, this method downplays or even ignores the less common, the less likely, the out-of-the-ordinary. Content analysis may often prevent us from thinking "out of the box". Let's face it, how many of us "googlers" go down the whole list and look at lower-matching hits? Terrorists who communicate about a planned attack without using recognized vocabulary may not be discovered by the global intelligence and surveillance systems. Market opportunities might be missed simply because analysts failed to use keywords corresponding to the words used by potential customers.

And you might miss an opportunity to get a cheap ticket to that remote location simply because you did not check alternative routes, airlines, dates, or means of transportation. Yet, whether we like it or not, online content analysis has become an almost indispensable part of our daily lives.

References

Adevaa, J. J. G., & Atxa, J. M. P. (2007). Intrusion detection in web applications using text mining. *Engineering Applications of Artificial Intelligence, 20*(4), 555–566.

Amoroso, E. G. (2011). *Cyber attacks: protecting national infrastructure*. Burlington: Elsevier.

Berry, M. W. (Ed.) (2004). *Survey of text mining: clustering, classification, and retrieval scanned by velocity*. New York: Springer.

Benford, R. D., & Snow, D. A. (1992). Framing process and social movement: an overview and assessment. *Annual Review of Sociology, 26*, 611–639.

Berry, M. W., & Kogan, J. (2010). *Text mining: applications and theory*. Chichester: Wiley.

Brabazon, T. (2006). The Google effect: googling, blogging, wikis and the flattening of expertise. *Libri, 56*, 157–167.

Bueno, P., et al. (2011). McAfee Threats Report: First Quarter 2011, MacAfee.

Carmines, E. G., & Zeller, R. A. (1979). *Reliability and validity assessment*. London: Sage.

Casey, E. (2011). *Digital evidence and computer crime: forensic science, computers and the Internet* (3rd ed.). Waltham/London: Academic Press/Elsevier.

Cordesman, A. (2002). *Cyber-threats, information warfare, and critical infrastructure protection: defending the U.S. homeland*. Westport: Praeger.

Druckman, D. (2005). *Doing research: methods of inquiry for conflict analysis*. London: Sage.

Entman, M. (1993). Framing: toward clarification of a fractured paradigm. *Journal of Communication, 43*(4), 51–58.

Hearst, M. A. (1999). Untangling text data mining. In *ACL '99 Proceedings of the 37th annual meeting of the association for computational linguistics on computational linguistics*. Stroudsburg: Association for Computational Linguistics.

Holsti, O. R. (1969). *Content analysis for the social sciences and humanities*. Reading: Addison–Wesley.

Janasik, N., Honkela, T., & Bruun, H. (2009). Text mining in qualitative research application of an unsupervised learning method. *Organizational Research Methods, 12*(3), 436–460.

Krippendorff, K. (1980). *Content analysis: an introduction to its methodology*. London: Sage.

Klotz, A., & Lynch, C. (2007). *Strategies for research in constructivist international relations*. Armonk: M.E. Sharp.

Kurtz, H. (2002). Robotic journalism: Google introduces Human-Less news. *The Washington Post*. 30 September 2002. Available at http://andrewcoile.com/CSUMB/2002/fall/CST373/scrapbook/robotjournalism.pdf. Last accessed 22 December 2011.

Lagerkvist, J. (2011). Inside the authoritarian state: new media entrepreneurs in China: allies of the party-state or civil society? *Columbia Journal of International Affairs, 65*(1).

Liu, B. (2011). *Web data mining: exploring hyperlinks, contents, and usage data* (2nd ed.). Heidelberg: Springer.

McGinty, S. (1999). *Gatekeepers of knowledge: journal editors in the sciences and the social sciences*. Westport: Bergin and Garvey.

Phillips, N., & Hardy, C. (2002). *Discourse analysis*. Thousand Oaks: Sage.

Srivastava, A. N. & Sahami, M. (Eds.) (2009). *Text mining classification, clustering, and applications*. London: CRC Press.

StatSoft, Inc. (2011). *Electronic statistics textbook*. Tulsa: StatSoft. Available at www.statsoft.com/textbook. Last accessed 22 December 2011.

Trochim, W. M. (1999). *The research methods knowledge base* (2nd ed.). Ithaca: Cornell University Custom Publishing.

Tsai, F. S., & Chan, K. L. (2007). *Threats in weblogs using probabilistic models. Lecture notes in computer science* (pp. 46–57). Detecting Cyber Security, 4430/2007.

Weare, C., & Lin, W. Y. (2000). Content analysis of the world wide web: opportunities and challenges. *Social Science Computer Review, 18*(3), 272–292.

Weber, R. P. (1990). *Basic content analysis*. London: Sage.

Weimann, G. (2010). *Terrorism and counterterrorism on the Internet. The international studies encyclopedia*. Denmark: Blackwell. Blackwell Reference Online 28 December. Available from http://www.isacompendium.com/subscriber/tocnode?id=g9781444336597_chunk_g978144433659719_ss1-18. Last accessed 22 December 2011.

Zheng, R., et al. (2003). Authorship analysis in cybercrime investigation. *Intelligence and Security Informatics, 2665*, 59–73.

Chapter 7
Secure Products Using Inherent Features

M. Blankenburg, C. Horn, and J. Krüger

Abstract It is a fact that counterfeiting jeopardizes the success of companies all over the world through violating intellectual property rights and causing enormous economic damage. This is because the counterfeit itself is mostly sold at a much lower price than the brand product. Some reasons could be the lack of development costs, a poor and cheaper quality of used materials or production and lower costs for mass production. Therefore it has become necessary for companies to secure their brands against counterfeiting.

Existing technologies for automatic fraud detection include additional security mechanisms like Data Matrix Codes or RFIDs added to the brand product itself which raises again the costs of the product. This text shows a new approach to secure a brand product by detecting product inherent features gained through the production process and the used materials. Therefore this new approach does not need additional features, nullifying this costs.

7.1 Introduction

The OECD's report on the Economic Impact of Counterfeiting and Piracy estimates a total loss of 250 billion dollars in (the year) 2007. This (economic) damage affects in particular countries that use advanced production and manufacturing processes based on intensive research and development to produce high quality goods. The review of copyright infringement of registered trademarks and products is not easy

M. Blankenburg · C. Horn
Technische Universität Berlin, Pascalstr. 8-9, 10587 Berlin, Germany

M. Blankenburg
e-mail: blankenburg@iwf.tu-berlin.de

C. Horn
e-mail: horn@iwf.tu-berlin.de

J. Krüger (✉)
Fraunhofer Institute for Production Systems and Design Technology, Pascalstr. 8-9, 10587 Berlin, Germany
e-mail: joerg.krueger@ipk.fraunhofer.de

J. Krüger et al. (eds.), *The Secure Information Society*, DOI 10.1007/978-1-4471-4763-3_7,   149
© Springer-Verlag London 2013

to implement. Due to the high number of pending trademarks and constantly added new applications it is very difficult for the executive bodies, such as customs, to register violations of trademark rights immediately and in a comprehensive manner. The awareness to all registered brands and products is for the executive organs not possible and therefore necessarily, trademark infringement remains unnoticed. The way to secure top-quality counterfeit products is often followed by an application of artificial security features. The issues of such security labels are in part the high cost, and additionally the integration into the product. High-quality branded products, as the target of counterfeiting, have usually, due to the production processes and materials used, and in view of its processing machinery and equipment, a grade of high quality. The specific conditions of production, manufacturing technologies and materials generate specific features, which identify the product uniquely. These features may be detected multimodal by man, including tactile (plasticity, elasticity, thermal conductivity, surface structure), visual (shape, colour, surface texture, transparency), olfactory (smell) or acoustic (sound) perceptions. In general, only the person familiar with the manufacture of the product can combine these inherent characteristics in their entirety so that it can differentiate the genuine product from a clear counterfeit. In the project Inherent-ID two properties of a product have been identified as the most promising ones suitable for identification: the olfactory and the optical features.

7.2 Motivation

The OECD report "The Economic Impact of counterfeiting and piracy" (OECD 2008) covers the analysis of international trade in counterfeit and pirated products. It estimates a trading volume of up to 200 billion USD in 2005 and up to 250 billion USD in 2007. These estimates do not include domestically produced and consumed counterfeit and pirated products and the significant volume of pirated digital products being distributed via the Internet. If these were also considered, the magnitude of counterfeiting and piracy worldwide could be several hundred billion dollars more than previously thought, and this increasing trend is quite alarming. It is self-evident that counterfeiting and piracy are businesses from which criminal networks thrive. The report shows further that the items counterfeiters and pirates produce and distribute are often of minor quality and can even be dangerous and health hazards. The effect of counterfeiting and piracy is an intermission of innovation and thus impairment of economic growth. With the magnitude of counterfeiting and piracy in mind, the report emphasizes the need for more effective enforcement to combat the counterfeiting and piracy on the part of governments and businesses alike. A key component for this enforcement is the development of methods for automated counterfeit detection.

123456-789

(a) Barcode   (b) QR-Code   (c) Other Data
                                                              Matrix Code

Fig. 7.1 Different types of codes

7.3 An Overview of Automated Counterfeit Detection Methods

Common automated counterfeit detection methods require nowadays additional security features at the product itself. Several methods have been developed, but main advantages and disadvantages remain similar.

Additional security features require further steps in production to add these features to the product. This raises expenses, manufacturing time and development efforts, which is clearly a disadvantage. On the other hand the security is enhanced and an original brand is easy to detect in an automated fashion, since there is a specific feature to look for. But this could also be a main disadvantage, if the security feature itself is easy to reproduce and could be added to any forged product. Figure 7.1 shows examples of different Data Matrix Codes which are commonly used on products for different purposes. One purpose is the use as a logical security feature where the printed security pattern contains unique information and cannot be copied.

Counterfeit detection without artificial security tags is a solution to these problems, if the counterfeit is distinguishable from the original brand.

7.3.1 Artificial Security Tags

The Anthology "Identification technologies to provide effective protection against product piracy" (Abramovici et al. 2010) gives a comprehensive overview of the latest efforts in product protection. A reasonably well studied approach is the extensive supervision of supply-chains. Here the application of RFID tags plays a significant role, as the latest form of artificial security tags, which can easily be integrated with existing logistic chains. The application of Data Matrix Codes (DMC) is discussed as well as a cost-effective alternative. Much work has been done to link these tags inseparably with the corresponding product to hinder product pirates from transferring these tags to their counterfeits. But in general it is observed that this protection method holds only with tremendous logistic implications, since todays products cover various stations during the distribution process. Up to now there has been no common standard available and the customs authorities' integration is still open. Even when the cost of these artificial tags could be reduced by advances in the production process, as e.g. the introduced direct printing of RFID antennas onto packaging, additional expenses with no direct use for the customer will arise. Security Tags like holograms found attached to various consumer goods give nearly no protection against counterfeiting since machine readability is poor and knowledge of the correct appearance is scarce.

7.3.2 Usage of Product-Inherent Features

In contrast to the usage of artificial security tags the project Inherent-ID (Krüger and Blankenburg 2010), initiated by the Fraunhofer Cluster of Innovation 'Secure Identity Berlin-Brandenburg', is elaborated by the Department of Industrial Automation Technology at Technische Universität Berlin. This project adopts a novel approach to protect high-value products from counterfeiting. The approach is based on the stationary and mobile capture of key product features indissolubly linked with the product which enable its production process to be traced. This not only renders the application of security tags obsolete but also gives enhanced protection against counterfeiting as the inherent characteristics cannot be removed from the product.

The Project Inherent-ID aims to answer the question: Which inherent features allow separation of genuine products from counterfeits in an automated fashion? The motivation of this question is the thesis that genuine products must differ in its properties from its counterfeit, since the product pirate tries to maximize its profit by using material of inferior quality and misusing a trademark of a genuine manufacturer to feint the customer. One result of the project is that only a combination of features can detect counterfeits at a decent rate for different products.

7.4 The New Counterfeit Detection Approach in Detail

Optical 2D and 3D characteristics as well as olfactory characteristics, which the high-quality production process impregnate in the genuine product, are combined with one another to serve as proof of product identity. They form the basis on which electronic certificates of authenticity can be issued without the need for complicated explicit security markings. Methods for the capture and control of identity characteristics are elaborated in the Inherent ID project for system integration using intelligent cameras and an electronic nose. The identity characteristics captured by this range of sensors serve both for product identification and product authentication. At the same time this also offers opportunities for improving documentation of product flows in the supply chain. Full documentation serves as a complement to the inherent characteristics of the authentic product and offers valuable information for verification of the genuine article, thus serving to safeguard against counterfeits. Within the scope of Inherent ID is the successful establishment of a laboratory providing multi-modal measurement equipment comprising multigas sensor array for olfactory analysis, high resolution camera for texture analysis and stereo vision, as well as range cameras for 3D feature extraction. Further research is conducted with the aim for increasing robustness of the sole test methods especially under ambiguous environments, integration into portable devices, realizing sensor data fusion for increased detection ratio, effortless integration into supply chains and developing efficient data models for storage of various features depending on the regarded product.

7.4.1 Texture

The ability to characterize visual textures and extract the features inherent to them is considered to be a powerful tool and has many relevant applications. A textural signature capable of capturing these features, and in particular capable of coping with various changes in the environment would be highly suited to describing and recognizing image textures (Xu et al. 2009). As humans, we are able to recognize texture intuitively. However, in the application of Computer Vision it is incredibly difficult to define how one texture differs from another. In order to understand, and manipulate textural image data, it is important to define what texture is. Image texture is defined as a function of the spatial variation of pixel intensities (Tuceryan and Jain 2001). Furthermore, the mathematical description of image texture should incorporate, identify and define the textural features that intuitively allow humans to differentiate between different textures. Numerous methods have been designed, which in the past have commonly utilized statistical models, however most of them are sensitive to changes in viewpoint and illumination conditions (Xu et al. 2009). For the purposes of mobile counterfeit detection, it is clear that this would be an important characteristic for the signature to have, as these conditions can not be entirely controlled.

Recently a description method based on fractal geometry known as the multifractal spectrum has grown in popularity and is now considered to be a useful tool in characterizing image texture. One of the most significant advantages is that the multifractal spectrum is invariant to the bi-Lipschitz transform, which is a very general transform that includes perspective and texture surface deformations (Xu et al. 2009).

Another advantage of Multifractal Spectra is that it has low dimension and is very efficient to compute (Xu et al. 2009) in comparison to other methods which achieve invariance to viewpoint and illumination changes such as those detailed in Varma and Zisserman (2002, 2003). One of the key advantages of multifractal spectra, which is utilized here is that they can be defined by many different categorizations or measures, which means that multiple spectra can be produced for the same image.

This is achieved through the use of filtering, whereby certain filters are applied to enhance certain aspects of the texture, to create a new measure. Certain measures are more or less invariant to certain transforms, and the combination of a number of spectra achieves a greater robustness to these. The workflow is depicted in Fig. 7.2 and an example is given in Fig. 7.3.

Fig. 7.2 Process chain for
the texture analysis

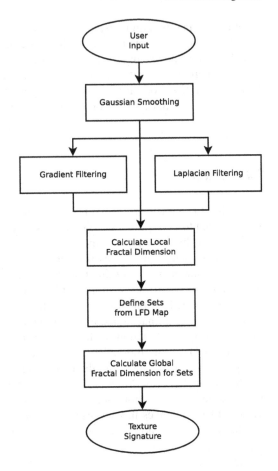

7.4.2 *Visual Object Recognition Using Shape*

Since manual detection is often done visual by customs officials, visual features
are also important for any automatic detection mechanism. Besides detecting features through two dimensional image processing, three dimensional data capture is
necessary for counterfeit detection, because it provides important additional information. To capture a real-world object in three dimensions a 3D scanner, or range
camera, can be used. The basic principles of 3D scanners available on the market
are triangulation, time-of-flight or interferometric approaches, whereas each principle has its advantages or disadvantages. For a profound insight into that topic refer
to Jähne (2005). We use a mobile structured-light 3D scanner for our application,
but in general any three dimensional data acquisition method can be used to capture
a real-world object. But using different kinds of scanning techniques results may
vary.

   One distinguishable feature of brand products is the shape itself. Shape matching
is a well studied topic and several publications can be found over the last 15 years.

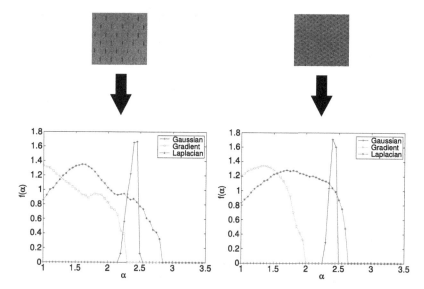

Fig. 7.3 Multi-Fractal-Spectra of texture of a textile product (*top*) and its counterfeit

Feature-based approaches have become very popular since some years in image analysis (2D) due to robustness and less computational effort compared to other approaches. In shape matching (3D) feature-based approaches have been introduced more recently and are gaining popularity in shape retrieval applications for the same reasons. The major difference is whether the approach uses global or local features. In Tangelder and Veltkamp (2007) an overview of shape matching principles and algorithms can be found.

Many shape matching approaches use digital human made data like the Princeton-Shape-Benchmark (Shilane et al. 2004) or the SHREC datasets (Bronstein et al. 2010a, 2010b, 2010c) to evaluate their algorithms. Scanned data from real world objects is different in a sense that holes[1] and variations between two scans of the same object can appear.

For that reason most approaches are not suitable for counterfeit detection, where minor details of an object can be highly important. For that reason only approaches detecting local features were taken into consideration. Figure 7.4 shows the required steps for our shape matching algorithm using real world objects. The shape matching algorithm requires a three dimensional model of the product as input which can be matched to an abstract model of the brand product. The abstract model is a description of features that render the brand unique.

One major challenge for three dimensional object capture is the huge amount of data that has to be processed. The 3D scanner we use has an accuracy of 20 to 50 μm and generates around 300,000 vertices per object. Assuming a point per point

---

[1]Holes are areas on the scanned object where the used scanning technique has troubles to capture data.

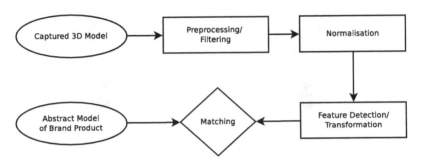

Fig. 7.4 Shape matching algorithm in counterfeit detection

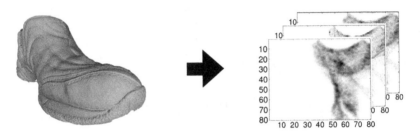

Fig. 7.5 Transformation of shape features

matching algorithm with $O(n^c)$ and $c > 1$ growth rate and a calculation time of 1 ms per point match, it would take nearly 3 years to calculate a match of two objects. This simple example demonstrates the challenge. Optimized algorithms, data reduction, parallel processing or transformation is necessary to achieve acceptable results.

For that reason the concept of *Key-Points* or *Points-Of-Interest* in combination with transformation is used. To do so, a feature detector (Bronstein et al. 2010a) has to be applied and the area surrounding the detected Key-Points is transformed into a meaningful descriptor.

Figure 7.5 shows a transformation of the area surrounding Key-Points into a 2D dense map using Spin Images (Johnson and Hebert 1999). A set of Spin Images is then transformed into a description of the object that can be matched to the abstract brand model.

7.4.3 Odour sensing—electronic noses

Much effort has been spent on how odour could be measured. The European Standard EN-13725 (EN 13725 2003) defines a method for the objective determination of the odour concentration of a gaseous sample using so called dynamic olfactometry. It is currently the only standardized method for the evaluation of odour impressions.

The dynamic olfactometry is a method where a panel of human assessors evaluates the concentration of odour in a series of standardized presentations of a gas sample. Here the emission rate of odours emanating from point sources, area sources with outward flow and area sources without outward flow are considered. The primary application of this standard is to provide a common basis for evaluation of odorant emissions in the member states of the European Union. Every method claiming the ability to detect arbitrary odour emissions has to benchmark against this standard. An overview of the development and application of electronic noses is given in Gardner and Bartlett (1999).

In general it was observed that electronic noses do not react to human inodorous gases and were also unable to detect some gases humans are able to smell naturally. Beginning with the working principle of specific gas sensors the concept of electronic noses as a combination of sensor array and diverse pattern recognition algorithms for classification is introduced. In principle the sensor concepts could be divided into three categories. The commercially available electronic nose Artinos basing on the KAMINA (KArlsuher MIkroNase) (Haeringer and Goschnick 2008) is a representative of metal conductance sensors. Here the sample gas flowing alongside the sensor surface is changing the concentration and configuration of oxide containing compounds, thus changing the conductance of the metal-oxide, which is then used as a measurement signal. The sensor elements differ by the thickness of silicon dioxide coating. Additionally the temperature is changed over time producing 38 analogue channels containing also transient responses, which are to be analysed. Due to its working principle these sensors deliver the most unspecific data, which is both an advantage and a disadvantage at the same time, since the sensors are suitable for a broad variety of samples, but the signal processing is harder to realize. A metal-oxide conductance sensor using 16 channels was utilized in the project Inherent-ID (Krüger and Blankenburg 2010). A typical pattern is shown in Fig. 7.6.

A similar sensor setup is used in Chilo et al. (2009), the difference being that the sensor elements are coated with different polymers, which induce a change in conductance to specific gas components. It was shown that with four different sensor types held at four different temperatures, so a total of 16 channels and following linear discriminant analysis ovarian cancer could be detected from tissue samples. There are still some issues with falsely rejected samples, but the results were quite impressive with respect to the use of ad-hoc methods. Another sensor concept utilizing polymer coatings are the quartz microbalance sensor arrays as described in Yuwono et al. (2003). These sensors detect the change of frequency when a gas is flowing over the sensor surface. In principle these arrays are very sensitive but also very susceptible to disturbances. Most of recently published results in odour detection are based on linear discriminant analysis and derivatives thereof. These methods are efficient in classification of complex sensor data, but with a manageable number of classes. And these methods need a significant amount of data present and are therefore not suitable for the here elaborated problem of one to many matching, as needed for the application in counterfeit detection. Instead effort is made in the extraction of relevant features for the

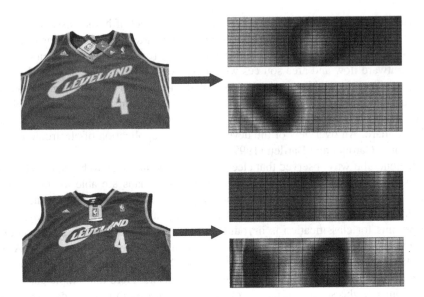

Fig. 7.6 Olfactory pattern of a genuine jersey (*top*) and a counterfeit (*bottom*)

purpose of reducing the dimensionality of the matching problem. An attempt of designing a general odour model was made in Bitter (2009), but was not successful due to the sensors used and the fact that nonlinear behaviour was excluded in advance.

7.4.4 Workflow

With the features described above there is a strong basis for automated classification of patterns. The key point for a robust and reliable counterfeit detection is the combination of these features and additional user information with the aim to derive a decision whether the probe is likely to be a counterfeit. An advantage of the proposed algorithms for feature extraction is the possibility to utilize statistical frameworks since the features are represented by probability distributions.

In general there are various approaches possible. Starting with a direct fusion of the features as proposed in Mitchell (2007) and shown in Fig. 7.7, or a more sophisticated approach which is taking the process of probing into account. Such a workflow is depicted in Fig. 7.8.

Here the decision process is not necessarily based on the utilization of all features, since some of them are dispensible or could be misleading. Think of the probing of shirt, obviously the 3D geometry cannot give a relevant contribution to the decision process and the 3D scanning can therefore be omitted. The classification itself is done with an adjusted Bayesian approach where special account was given

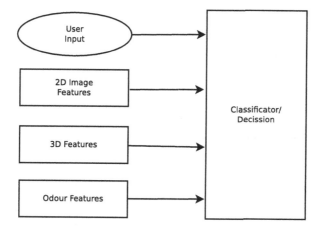

Fig. 7.7 Feature fusion concept

to the detection of novel and therefore unknown patterns. This was done with estimation of the Level of Significance distribution, which gives additional to a decision information an value of the plausibility of this decision, cf. Kühn (2010).

7.5 Conclusion

It was shown that the Inherent-ID Project adopts a novel approach to protecting high-value products from counterfeiting. The approach is based on the stationary and mobile capture of key product features indissolubly linked with the product which enable its production process to be traced. This not only renders obsolete the application of security tags but also gives enhanced protection against counterfeiting as the inherent characteristics that the high-quality production process impregnate in the genuine product are combined with one another to serve as proof of product identity. They form the basis on which electronic certificates of authenticity can be issued without the need for complicated explicit security markings. Methods for the capture and control of identity characteristics are being elaborated in the Inherent-ID project for system integration using intelligent cameras and an electronic nose. The identity characteristics captured by this range of sensors serve both for the product identification and product authentication. At the same time this also offers opportunities for improving documentation of product ows in the supply chain. Full documentation serves as a complement to the inherent characteristics of the authentic product and offers valuable information of verification of the genuine article, thus serving to safeguard against counterfeits.

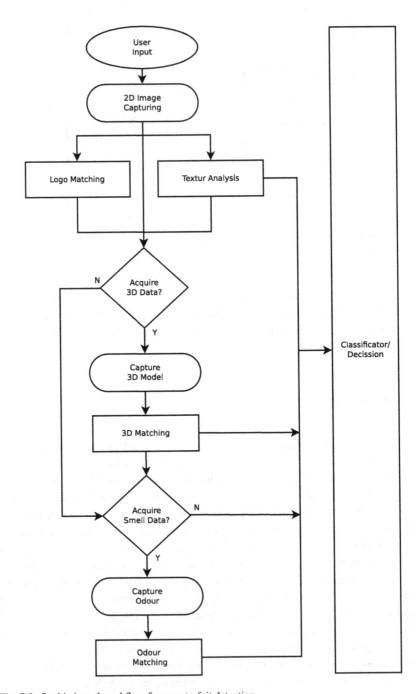

Fig. 7.8 Sophisticated workflow for counterfeit detection

Acknowledgements   The authors would like to acknowledge the funding of the research project Inherent-ID by the senate of the state Berlin and the European Regional Development Fund. The project is embedded in the Fraunhofer Cluster of Innovation Secure Identity.

References

Abramovici, M., Overmeyer, L., & Wirnitzer, B. (2010). *VDMA Innovations for anti-counterfeiting: Vol. 2. Identification technologies to provide effective protection against product piracy*.

Bitter, F. (2009). *Modell zur Bestimmung der Geruchsintensität der Raumluft mit Multigassensorsystemen*. Doctoral thesis, TU, Berlin.

Bronstein, A. M., Bronstein, M. M., Bustos, B., Castellani, U., Crisani, M., Falcidieno, B., Guibas, L. J., Kokkinos, I., Murino, V., Ovsjanikov, M., Patané, G., Sipiran, I., Spagnuolo, M., & Sun, J. (2010a). SHREC 2010: robust feature detection and description benchmark. In *Proc. EUROGRAPHICS workshop on 3D object retrieval (3DOR)*.

Bronstein, A. M., Bronstein, M. M., Castellani, U., Dubrovina, A., Guibas, L. J., Horaud, R. P., Kimmel, R., Knossow, D., von Lavante, E., Mateus, D., Ovsjanikov, M., & Sharma, A. (2010b). SHREC 2010: robust correspondence benchmark. In *Proc. EUROGRAPHICS workshop on 3D object retrieval (3DOR)*.

Bronstein, A. M., Bronstein, M. M., Castellani, U., Falcidieno, B., Fusiello, A., Godil, A., Guibas, L. J., Kokkinos, I., Lian, Z., Ovsjanikov, M., Patané, G., Spagnuolo, M., & Toldo, R. (2010c). SHREC 2010: robust large-scale shape retrieval benchmark. In *Proc. EUROGRAPHICS workshop on 3D object retrieval (3DOR)*.

Chilo, J., Horvath, G., Lindblad, T., & Olsson, R. (2009) *Lecture notes in computer science: Vol. 5633. Electronic nose ovarian carcinoma diagnosis based on machine learning algorithms*.

EN 13725 (2003). *Air quality. Determination of odour concentration by dynamic olfactometry*, DIN EN 13725:2003.

Gardner, J. W., & Bartlett, P. N. (1999). *Measurement science and technology: Vol. 11. Electronic noses—principles and applications*. Oxford: Oxford University Press.

Haeringer, D., & Goschnick, J. (2008). Characterization of smelling contaminations on textiles using a gradient microarray as an electronic nose. In *Sensors and actuators B-chemical* (Vol. 132), Nr. 2.

Jähne, B. (2005). *Digital image processing*. ISBN 3-540-24035-7.

Johnson, A., & Hebert, M. (1999). Using spin images for efficient object recognition in cluttered 3d scenes. *IEEE PAMI, 21*, 433–449.

Krüger, J., & Blankenburg, M. (2010). Secure identity—a source for innovative it-systems and processes. In *Future security*, 5th security research conference, Berlin, 7–9 September 2010. Freiburg: Fraunhofer VVS.

Kühn, S. (2010). *Stochastic Engineering – Berechnung, Entwicklung und Modellierung bei unsicherer Information*. Doctoral thesis, TU, Berlin.

Mitchell, H. B. (2007). *Multi-Sensor data fusion: an introduction*. Berlin: Springer.

OECD (2008). *The economic impact of counterfeiting and piracy*. Paris: OECD. www.oecd.org/sti/counterfeiting.

Shilane, P., Min, P., Kazhdan, M., & Funkhouser, T. (2004). *Shape modeling international. The Princeton shape benchmark*.

Tangelder, J. W., & Veltkamp, R. C. (2007). *A survey of content based 3D shape retrieval methods: Vol. 39. Multimedia tools and applications*. (pp. 441–471), No. 3.

Tuceryan, M., & Jain, A. K. (2001). *Texture analysis, handbook of pattern recognition & computer vision* (2nd ed.). Singapore: World Scientific

Varma, M., & Zisserman, A. (2002). Classifying images of materials: achieving viewpoint and illumination independence. *ECCV, 3*, 255–271.

Varma, M., & Zisserman, A. (2003). *CPVR: Vol. 2. Texture classification: are filter banks necessary?* (pp. 691–698).

Xu, Y., Ji, H., & Fermüller, C. (2009). Viewpoint invariant texture description using fractal analysis. *International Journal Computer Vision, 83*, 85–100.

Yuwono, A. S., Hamacher, T., Nieß, J., Boeker, P., & Lammers, P. S. (2003). Implementation of a quartz microbalance (QMB) sensor array—based instrument and olfactometer for monitoring the performance of an odour biofilter. In *2nd IWA international workshop & conference on odour & VOC's*, Singapore.

Chapter 8
Assistant-Based Reconstruction of Believed Destroyed Shredded Documents

D. Pöhler, J. Schneider, and B. Nickolay

Abstract This article introduces the assistance system for the virtual reconstruction of shredded documents developed at Fraunhofer IPK. The system enables the user to reconstruct documents, which often could not be reconstructed by hand. By this system the information contained in the documents can be made usable again.

The process of the assistance system is divided into the four phases of Digitalization, Feature-Extraction, Context Matching and Interactive Viewer.

In the first phase the fragile and often twisted strips to be reconstructed have to be separated and straightened. They then need to be scanned double-sided, anechoic, and shadeless as well as with highest colour and geometrical fidelity compared to the originals. In the second phase the strips are segmented pixel by pixel from the raw scans. Foreground from background information is separated and describing content features for a discriminative similarity comparison are extracted from the digitized strips. The third phase works on the basis of these calculated features. At first a search space reduction is carried out. Afterwards a pairwise context matching is executed by using dynamic programming algorithms. Thereby similarity scores of all possible combinations of shredder strip pairs are calculated. On the basis of these scores a semi-automatic reconstruction is performed in the interactive viewer in phase four. By doing so strip-pairs are put together gradually to grow partial reconstructions up to complete reconstructed documents by user interaction.

In the past, numerous enquiries within the framework of tax and murder investigations could be solved successfully by means of this assistance system, enabling the evaluation of procedural relevant documents; which had been destroyed purposely.

8.1 Introduction

In the last years the Fraunhofer Institute for Production Systems and Design Technology IPK has systematically developed and successfully implemented methods

D. Pöhler · J. Schneider · B. Nickolay (✉)
Fraunhofer Institute for Production Systems and Design Technology IPK, Pascalstrasse 8-9, 10587 Berlin, Germany
e-mail: bertram.nickolay@ipk.fraunhofer.de

J. Krüger et al. (eds.), *The Secure Information Society*, DOI 10.1007/978-1-4471-4763-3_8,   163
© Springer-Verlag London 2013

for the automated virtual reconstruction of destroyed or damaged documents. The virtual reconstruction is based on several algorithms of pattern recognition and digital image processing. With this interdisciplinary technology it is basically possible to digitize damaged documents and to merge the digital document fragments automatically. Thus, it enables different users to analyze and index data, i.e. to store the indexed data in an organized database appropriately.

One of the most well-known applications developed by Fraunhofer IPK is a technology, which could make a great contribution to the historical investigation and historical research, for forensic, as well as to the preservation of cultural heritage: the automated virtual reconstruction of torn, shredded and otherwise damaged documents and planar objects. By means of the automated virtual reconstruction, digital images of fragments can be virtually, computer-assisted reconstructed and e.g. documents become readable again, whose reconstruction by hand would be too work-intensive or simply not possible.

Currently, the emphasis of development at Fraunhofer IPK is placed on the realization of a continuous system for the reconstruction of documents of the state security service (*Stasi*) of the former German Democratic Republic (GDR). During the reunification of Germany in 1989/1990, this state security service has torn many documents into pieces. Now, these documents need to be reconstructed virtually in order to make them available to us and ensuing ages. Due to the enormous amount of fragmented documents—approximately 40 million pages of paper have been torn into about 600 million pieces—a reconstruction system with a high level of automation is indispensable.

The attained results of this development have been constructive and trend-setting not only for this scientific discipline, but also for disciplines of other fields, such as forensic science, archive studies, and antiquarian studies. Besides paper, for example, papyrus and vellum play for millennia an important role as media for culture and information preservation. Their damaging—accidently or intentionally—led to a loss of all the information they contained if the manual reconstruction was either outright impossible or would simply require too much expenditure of time or personnel.

Even though the system for the virtual reconstruction is mainly in use for the reconstruction of hand-torn and shredded paper, though the underlying methods were also successfully applied for the reconstruction of other plane objects, such as cut up number plates, in the past (see Fig. 8.1).

8.1.1 Overview of the Overall Process of Virtual Reconstruction

In general, the overall process of virtual reconstruction is divided into three process steps: (1) digitalization of fragments, (2) virtual reconstruction of digitalized objects ("ePuzzler"), as well as (3) processing and analysis of the individual objects to entire processes.

Fig. 8.1 Examples for virtual reconstructions. The figure shows complete reconstructions of lengthwise-cut shredded pages (*left* and *centre bottom*); partial reconstructions of a torn vehicle number (*centre, top*) and a cross-cut shredded DIN A4 page (*right*)

In a first step, digital images of the material, which needs to be reconstructed, have to be produced, i.e. fragments have to be scanned. Before scanning, it is necessary to go through some preparatory steps. It is for example crucial to separate the fragments, clean them, if necessary, straighten wavy or creased material, and remove foreign objects like paper clips or staples.

Subsequent to the digitalization, the second step, the real virtual reconstruction of damaged documents with the so-called "e-Puzzler" is performed. The ePuzzler is a modular, adaptive reconstruction software, which has been developed by Fraunhofer IPK. By means of complex image processing and pattern recognition algorithms, digital fragments are automated put together to complete objects by the ePuzzler. Hereby, the level of automation to be achieved strongly depends on properties and quality of the fragments, which have to be reconstructed. Furthermore, the ePuzzler provides the user with a wide range of tools, which differ depending on the particular task. With the help of these tools, questionable puzzle results can be checked manually or reconstruction proposals of the software puzzled interactively.

The third step goes beyond the reconstruction of individual objects and encompasses methods for the computer-aided formation of reconstructed objects to whole processes as well as the analysis of their content. These procedures are not considered any further in the following.

8.1.2 The "Core" of Virtual Reconstruction: The ePuzzler

The ePuzzler is the "core" of the system and is divided into three main components: feature extraction, search space reduction and matcher.

The methodology of the virtual reconstruction equals the one of a human being doing a jigsaw puzzle. He decides on the basis of numerous descriptive characteristics, if two fragments match or not. At first, analogue to the human approach, the ePuzzler calculates several different fragment features, e.g. texture characteristics, which are derived from the fragments' content, or characteristics of global or local colour value. These features are for instance used to reduce the combinatorial complexity of the actual puzzling. This is particularly important for large amounts of data. Hence, fragments, which are similar to each other regarding their features, are grouped by means of intelligent search space reduction and also summarized into sub-groups. Within these so-called search-space-sets, the actual reconstruction, the matching, takes place. Thereby, fragments are compared to each other regarding their feature consistency. If two fragments match, they are digitally merged, their common features re-calculated and considered as one larger fragment for further reconstruction.

8.1.2.1 ePuzzler and Expert Knowledge—Assistance Systems for Virtual Reconstruction

The automated virtual reconstruction is a very versatile tool. In order to apply this tool to specific tasks even more efficiently, Fraunhofer IPK is developing methods for assistance-based reconstruction systems, which make use of expert knowledge for the reconstruction process.

When talking about assistance-based reconstruction, we have to differentiate between two different approaches. On the one hand, there is the concept of assistance-based virtual reconstruction, on the other hand, the vision of an assistance system for the support of physical reconstruction.

Like for the automated virtual reconstruction, the aim of assistance-based virtual reconstruction is the *virtual* reconstruction of contents. The difference between both methods is: the assistance-based approach is designed for fragments, for which a fully automated reconstruction is not possible and the interactive participation of an expert functioning as a system operator becomes necessary. One example of such an assistance-based virtual reconstruction is the system for the reconstruction of shredded documents, which will be described later on.

Regarding an assistance system for the *physical* reconstruction, the outcome of the virtual reconstruction should serve as a master for the subsequent manual physical reconstruction. One example for such an application is the above-mentioned case of cut vehicle number plates, where the physical reconstruction of the original metal sheets has high priority. Such systems place high demands on the workflow before digitalization as well as on the operational procedure after the virtual reconstruction, especially if they are supposed to be laid-out for the processing of large amounts of data. Furthermore, the piecing together of the originals downstream of the virtual reconstruction requires the use of a high-capacity tracking system, which is adjusted specifically for the respective task.

8.1.3 Assistance System for the Reconstruction of Shredded Documents

Within the area of virtual reconstruction of two-dimensional objects, the reconstruction of shredded documents is a special case. In general, the paper strips, which have to be reconstructed, are only a few centimetres in length and a few millimetres in width. Hence, the methods of feature extraction and matching are confronted with great challenges. On the one hand, only a small amount of describing features may be deduced from the few pixels available. On the other hand, the strips have a uniform contour. Therefore, the reconstruction can solely be based on content features.

The digitalization of shredded paper strips is an enormous challenge as well as in general the objects to be scanned are extremely fragile, often twisted with each other, creased or bent. Ahead of digitalization, these objects have to be separated and straightened.

In the following sections, the technology for the assistance-based virtual reconstruction of shredded documents, which has been developed at Fraunhofer IPK, will be described. At first, in Sect. 8.2, essential requirements for the digitalization of paper strips are described. Subsequently, in Sect. 8.3, specific characteristics of the data material, which constitutes the basis of reconstruction, will be summarized. Section 8.4 contains a description of the implemented assistance system including its image-processing and pattern recognition methods, which have been applied successfully several times in the past.

8.2 Requirements for the Digitalization

The methods of digitalization of shredded paper strips have to meet high demands, because the high quality of the digitized images is a pre-condition for the precise and error-free reconstruction later on in the process. Scanners in use have to be able to scan strips of nearly every imaginable length—ranging from DIN A3 format longitudinal cut to only a few centimetres—, double-sided, anechoic, and shadeless as well as with highest colour and geometrical fidelity compared to the originals. Otherwise, fine correlations between strips belonging together, like e.g. identical colouring or texture, cannot be identified during the process of reconstruction. Furthermore, analogue to the blue screen technique known from movie technology, the strips have to be digitized against a background, which enables the pixel accurate masking out of the strips. Black or white backgrounds, like they are used in common scanners, are unsuitable, but e.g. lurid neon colours are appropriate. Moreover, especially in the case of large amounts of original material, a flow capacity of a few hundred to a few thousand objects per day may be aimed at, but the scanning material can be put a strain onto physically only minimally. Currently there is no scanner, which fulfils all these requirements. In cooperation with several scanner producers, the machines in use as well as the procedures applied at Fraunhofer are constantly

Fig. 8.2 "Initial situation" for the reconstruction of shredded documents

being adapted and further developed as regards to their use for the process of virtual reconstruction in particular.

One of the greatest challenges we are confronted with is the limitation of colour and geometrical deviation to a tolerable level. The deviations may be due to either the aging of the light source, the used scanning method or the application of multiple scanners. Colour differences can be corrected by means of colour management systems, whilst inaccuracy of the shape images is far more difficult to measure and correct.

Also the double-sided digitalization requires well-founded solutions. A definite attribution of paper front and back side is only possible if the orientation of both paper sides to each other is uniform during scanning. This can be reached by either simultaneously scanning both paper sides or using a feeding system, which ensures that the paper strips are fixed during turn-over. For those means, pockets of foil are especially qualified, because they are also of great value for the processing of tiny strips and the general protection against pressure and friction.

Additionally, besides the images of paper strips, meta-information may be collected during digitalization, which might be crucial for the reconstruction. If, for example several strips are stapled together, this information will be preserved and the reconstructed pages will be identified as belonging together. Furthermore, information about the rough classification with regards to context and time as well as storage location or place where it was found should be fixed in meta-data, if this information exists at all. By the preservation of this "previous knowledge", the performance of the puzzle process can be enhanced immensely and a context-specific evaluation of reconstructed objects will be eased.

In general, the separation of paper strips, which has to be carried out before digitalization, is very complex and, consequently, time-consuming. Figure 8.2 clearly shows: The individual strips are often twisted with each other in such a way, that a separation could only be performed with great manual effort up until now. Due to the high fragility of paper strips, the practicability of approaches with a high level of automation seems highly unlikely.

A general problem for the separation of paper strips, which form a big ball of paper strips as shown in Fig. 8.2, is the flexible characteristic of these strips. When breaking up such a ball of paper strips, there is always the risk of "balling" the paper strips further and benting, deforming or tearing the paper additionally.

Nonetheless, for an efficient batch processing, the strips have to be separated, straightened and fed to the scanning process much faster than it has been the case so far. Hence, future solutions are contemplated, which plan on a low level of automation for the scan preparation and feeding. These processes are carried out solely manually so far. By a low level of automation, at least supporting systems could be made available to the user. One possible solution for this complex of problems could be to feed the originals to a rotating roller, which has circumferentially positioned holes by means of a revolving screen or concial drum system. Through the rotation and the defined sizes of the holes, the pieces are being separated. This separation procedure is applied to numerous processes, e.g. for the separation of synthetics materials of different sizes from each other and to separate cardboard from paper.

8.3 Characteristics of the Data

The success rate of shredded document reconstruction strongly depends on the properties and condition of the paper as well as the construction type and condition of the paper shredder used. Depending on the paper material type and the sharpness of the paper shredder blades, the reconstructed strips present themselves quite differently. Thereby, the condition of cut edges varies from precise and sharply bounded to slightly frayed up to being in tatters, if the blades were blunt or the paper too compact. Especially in the latter case, strips produced with the very same paper shredder may show differing widths and do not necessarily have to be rectangular.

Furthermore, depending on the construction type of the paper shredder, the paper may either be cut into narrow strips (*longitudinal-cut*) or by means of additional horizontal cuts cut into tiny fragments (*cross-cut*).

Varying depending on the level of security of the used paper shredder, cross-cut shredded paper strips have a length of a few centimetres and a width of a few millimetres; at times the strips are even smaller than one millimetre. Moreover, cross-cut shredded material is often damaged much more than longitudinal-cut shredded material. These damages are mostly caused by a grinder, which works with rollers. Thereby, paper strips with very fuzzy running cut edges may be generated, which in turn leads in general to a digitized image with a high level of noise. That means, possible evaluable content of the strip edges are heavily disturbed by the grinding process of the paper shredder.

Longitudinal-cut strips are normally considerably larger than cross-cut strips. Depending on the security level of the paper shredder used, the width of the strips

Fig. 8.3 Classification of foreground objects. Not even a human examiner is in a position to classify the marked symbols on the two individual strips clearly (**a**). Only within the context of associated strips, the real meaning of these symbols becomes clear (**b**)

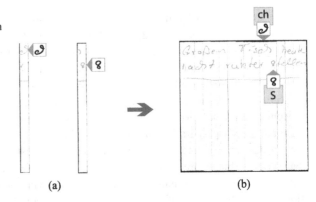

(a)   (b)

varies in most cases from two to ten millimetres. The length of the strips is given by the paper format of the original. Additionally, the length is determined by the angle in which the paper was fed to the paper shredder. Hence, longitudinal-cut shredded strips of one dataset, which needs to be reconstructed, usually show different lengths. If the paper was creased before cutting or the paper shredder had very blunt blades, "arched, waved, or zigzag strips" may be generated, too.

Besides others, all these circumstances can lead to reconstruction results showing document information like lettering, graphics, lining etc., which is staggered after digitalization, but was located right next to each other before shredding the material. Since this is the case for most actual applications and inquiries, which Fraunhofer IPK has been approached with in the past, it is no exception, but the rule. Hence, the comparing algorithms of the reconstruction system should in principal be able to work most efficiently with strips of differing lengths and match them with each other.

Independent of the cutting method of the paper shredder and the physical deformation of the cut data material, the variance of occurrence of information on the strips is extremely high in actual applications. This so-called foreground information or the amount of foreground objects derived from it respectively may not be classified accurately unless it is a matter of machine-generated text. As usually the strips are only a few millimetres wide, all content characteristics have to be derived from only a few pixels. For this reason, a robust classification, i.e. a reliable interpretation of foreground objects is in many cases hardly realizable, even by a human expert, if it is possible at all (cf. Fig. 8.3).

Therefore, the reconstruction system for shredded documents sets the classification of content into self-suggesting groups like "lettering" or "lining" aside. Instead, all elements on one strip—including paper colour—are treated as one geometric object. This leads to the situation that in the process of determining probable paper strip neighbours, only geometrical solutions suggest themselves, but not necessarily a content-wise correct result can be achieved. This is another reason why the reconstruction of shredded paper strips can only be carried out with an assistance system at the moment.

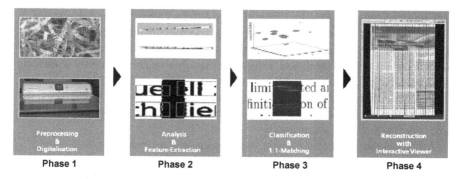

Fig. 8.4 Process chain of the virtual reconstruction of shredded documents

8.4 Virtual Reconstruction

The virtual reconstruction of shredded documents may be considered a linear process chain and is divided into four phases (see Fig. 8.4).

Each shredding strip has to pass through a continuous process ranging from scanning to partial to full reconstruction. Thereby, every individual strip is stored in a database, where it is clearly identifiable and a status is added to it. During the overall process the strips go successively through the following status: 'scanned & checked' (Phase 1), 'segmented & features calculated' (Phase 2), 'forming search spaces & carrying out pairwise matching' (Phase 3) as well as 'reconstructed' (Phase 4).

Phase 1—Digitalization   In this phase, paper strips are digitized from both sides with a resolution of 300 dpi and a colour depth of 24 bit, taking into consideration the general requirements described in Sect. 8.2. The outcome of this phase is a double-sided so-called raw scan with—depending on the scope of work—one or more digitized images of the paper strips in front of a neon-coloured, in general light green, background.

Phase 2—Feature Extraction   The first step of this phase is the cutting out of the individual shredding strips from the raw scan, i.e. to segment them pixel by pixel. On the basis of their edges, the segmented strips are now being erected with ±180° preciseness and saved in a database. After this process step, two digitized images of every single physical shredder strip are available within the reconstruction system: one *strip image* of the front and one of the back side of the paper strip.

The next step of this phase is the extraction of describing features (see Sect. 8.4.1), which have to be taken into consideration for the classification of strip images. Given that the reconstruction of shredded documents can only take place on the basis of content characteristics, the automated separation of strip images with and without content is of great importance within this step. The strips without content must be rejected, thereby diminishing the "candidates" for reconstruction in this early stage of reconstruction already. The rejected shredder strips may be from the border of or the blank back sides of the documents. In the first case, the shredder strips may not be used at all and have to be filed in a separate category within the database labelled 'not reconstructable' and be excluded from the further reconstruction process. In the latter case, the corresponding paper side will be tagged as 'not reconstructable' and the system will only use it for reconstruction proposals of back sides during the further reconstruction process.

All processing steps of this phase are running fully automatic.

Phase 3—Context Matching   After describing features of strip images have been calculated in phase 2, in phase 3 a so-called search space reduction and context matching of strip images within the obtained subsets using the calculated characteristics are automatically executed (see Sect. 8.4.2). Context matching within the subset constitutes the terminal step of the automatic process chain. It consists of the identification of similarity scores—so-called matching scores—of all possible combinations of shredder strip pairs.

Phase 4—Interactive Viewer   Within this phase the real reconstruction of shredded documents is carried out with the help of an interactive component (or: *interactive viewer, IAV*). On the basis of pairwise matching scores calculated during context matching, *IAV* proposes the most probable candidates for strip combinations to the user. Thereby, the gradual reconstruction of shredded strips to a complete document can take place. For this purpose, the user has to initiate several processes when using the *IAV*, like e.g. the digital merging of matching shredder strips, the update of the database after a successful merge or the rotation of "reconstruction candidates" (see Sect. 8.4.3).

8.4.1 Feature Extraction

With the end of the first processing step of phase 2, the images of the shredder strips to be constructed are available in the database segmented pixel by pixel and erected with $\pm 180°$ preciseness. They are erected in such a way, that the *cut edges* of the strips created by the shredder blades are vertically and the assumed *outer edges* of the original piece of paper are nearly horizontally aligned—depending on the angle in which the original document was shredded.

In a next step, the erected colour images of strips are binarized with the help of a quantization method. Hereby, a binary image is created in which the image foreground, e.g. a label, is separated from the image background, which is in general the paper colour.

Before the extraction of features can take place, artefacts which are smaller than the predetermined minimal size are excluded from the binary image by means of a modified Hit-or-Miss Transformation (HMT). What we call artefacts in this context

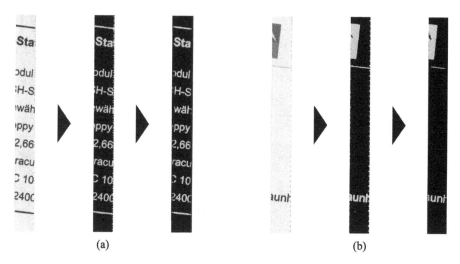

Fig. 8.5 Feature extraction. (**a**) and (**b**) each show, from *left* to *right* respectively: segmented shredder strips, the binary image derived from it, and a binary image filtered by means of HMT. (The *small dots* in the images in the middle are "artefacts" along the cut edges, which are filtered out by HMT)

are a few coherent pixels along the cut edges, which could be detected as foreground by mistake, but are only frayed cut edges caused by blunt blades of the shredder (see Fig. 8.5).

At first, we have to define two areas within each strip image for feature extraction: one along the left, one along the right cut edge of the strip. The definition of these cut edge areas takes place dynamically and depends on the strip width as well as on the run of the outer edges of the strip.

Two so-called feature strings per cut edge area are calculated: one *colour feature string* and one *object feature string*. The colour feature string is extracted from the cut edge area of the colour image of the strip, thereby describing the colour gradient along the cut edge. The object feature string is extracted from the binary image, thereby describing the allocation of foreground information along the cut edge. Hence, for each strip image, two colour and object feature strings are calculated, which sums up to four colour and object strings for each individual shredder strip.

These feature strings form the basis for the derivation of context representation.

The main idea of context representation is to summarize pixel lines with a similar distribution of foreground information and thereby describing possible coherent objects, like e.g. lines or paragraphs.

Furthermore, so-called Local Points of Interest (LPOIs) are detected on the basis of previously calculated object feature strings (see Fig. 8.6). These LPOIs are represented by the involved substrings of object and colour feature strings respectively.

All LPOIs together form the basis for context matching.

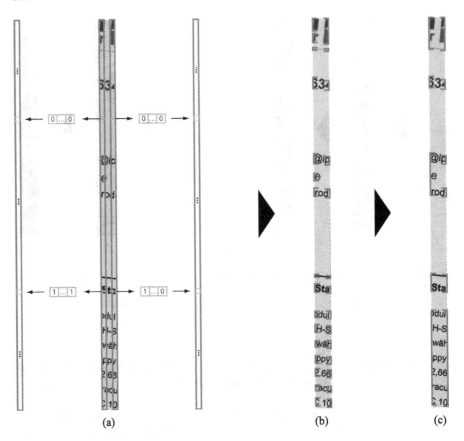

Fig. 8.6 Detection of local points of interest (LPOIs). Object feature strings are extracted for each cut edge area (**a**), individual lines on these strings are detected (**b**), and these lines are combined to paragraphs (**c**). The marked sections (**c**) are LPOIs

8.4.2 Context Matching

Aim of context matching is the determination of a similarity value, a so-called *similarity score* for a physical shredder strip pair on the basis of its content characteristic.

The input of the context matcher are four strip images of two physical shredder strips as well as their context representation in the form of the previously calculated LPOIs. With this input, the context matcher detects local optima for each possible cut edge pair by using dynamic programming algorithms. On the basis of these optima a global scoring in relation to cut edge pair is carried out.

The result of this algorithm is a similarity value, which belongs to the most compatible cut edge combination as well as to the corresponding precise vertical orientation of these strips to each other.

In general, for the matching of a cut edge pair applies: it should not be looked for identical foreground information on both strips. More than that one has to hypothe-

Fig. 8.7 16 possible cut edge pairs for each pair of physical strips

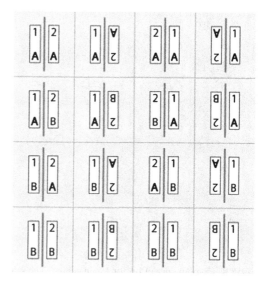

size that strips belonging together may only show a certain amount of continuation of foreground information and this information can only show a certain amount of similarity to each other. Consequently, the demand for a 100 % correspondence of two compared object feature strings of data material would not be very useful, unless the data material has been produced artificially. Additionally, strips belonging to each other may show for example inconsistent colours on the cut edges due to different effects of light or differing levels of damage caused e.g. by water. That's why the procedure of matching colour feature strings is designed even more tolerant than the matching of object feature strings.

When matching two physical shredder strips, all in all 16 so-called *1:1-matchings* of cut edges are performed, because all possible cut edge pairs of the four strip images involved have to be taken into consideration and evaluated in relation to each other.

If the four strip images of front and back side of a physical matching (1, 2) are named 1A, 1B, 2A, 2B, taking into consideration the correlation of side A and B of the strips, the four strip images may be combined as follows: 1A:2A & 1B:2B, 1A:2B & 1B:2A. For every singly one of these four combinations, the following four combinations of cut edge pairs have to be taken into consideration: 1R:2L, 1R:2R$_{180°}$, 1L:2R and 1L:2L$_{180°}$, whereby R and L stand for the right or left cut edge respectively and 180° for a rotation of 180° of the according cut edge and the corresponding feature vectors (see Fig. 8.7).

During *1:1-Matching* the optimal alignment and a score for each of the 16 possible pairings of the cut edges are calculated on basis of the previously calculated LPOI according to the same principle.

Due to the combinatory effort of the determination of similarity scores by pairs, this calculation constitutes the bottleneck of the overall system (see Fig. 8.8). The combinatory effort for all *1:1-matchings* to be carried out for n physical strips is

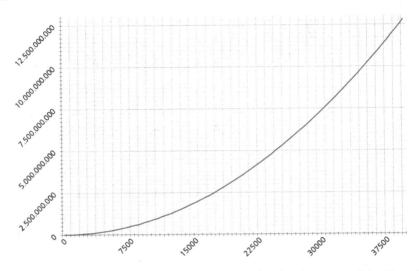

Fig. 8.8 Combinatory effort for 1:1-matching. The x-axis describes the number of shredder strips to be processed, the y-axis the number of possible cut edge pairs, i.e. the number of 1:1-matchings to be carried out

$\binom{n}{k} * 16$, whilst $k = 2$, because pairs are formed and 16 cut edge combinations are possible for each pair of physical strips.

8.4.2.1 Distinction of Cases *Aligned* and *Not-aligned*

Shredder strips do not necessarily need to fit to each other flush and even longitudinal-cut strips of one piece of paper may show different lengths. Therefore, during *1:1-matching*, we have to generally differentiate between the two matching situations *aligned* and *not-aligned* on the basis of length differences of cut edges in comparison to each other or the corresponding feature vectors respectively.

When talking about the matching situation *aligned*, it is assumed that two cut edges in comparison to each other can be fixed to their outer edges vertically. Therefore, when determining the scores of such cut edge combinations, only those LPOIs of both cut edges have to be compared to each other in pairs, which show overlapping y-coordinates.

When talking about the matching situation *not-aligned*, it cannot be assumed that two cut edges in comparison to each other can be fixed to their upper outer edges vertically, because the cut edges are slidable towards each other vertically. Depending on the length difference of both cut edges, an according number of orientation possibilities exists. For the determination of "meaningful" orientations, only the *LPOIs* of both cut edges are compared to each other in pairs. This way, the number of possible orientations is reduced and the level of performance of the *1:1-matching* boosted significantly.

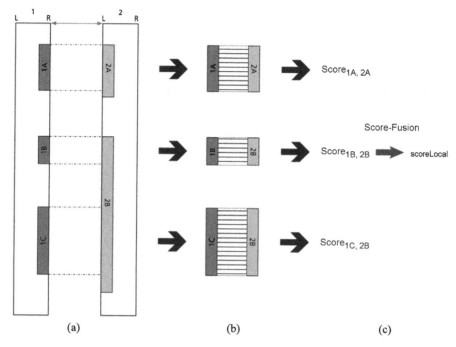

(a)   (b)   (c)

Fig. 8.9 Step *A1* of the matching situation *aligned*. For this step, the cut edges (*1R*, *2L*) are fixed flush vertically at their upper outer edges and the LPOIs of *strips 1* (*1A*, *1B*, and *1C*) as well as *strips 2* (*2A* and *2B*) are cut according to their vertical orientation (**a**). Subsequently, for every cut area a score is calculated (**b**) and the scores are fused to the similarity value *scoreLocal* (**c**)

8.4.2.2 1:1-Matching Situation *Aligned*

The *1:1-matching* in the case of *aligned* is divided into two steps: (A1) determination of local optima and (A2) execution of a global scoring on the basis of local optima.

During step (A1), the cut edges are fixed vertically at their upper outer edges. The whole vertical section, which is being involved by both parts of the cut edge pair, is called *matching section*. Within this matching section, the LPOIs of the cut edges are cut in order to define the sections involved conjointly. These common areas are being matched to each other by means of dynamic programming algorithms and the results are fused to the local similarity value *scoreLocal* (see Fig. 8.9).

In step (A2) a global scoring is executed by evaluating mathematically the different local parts, which were generated when cutting the *LPOIs* along the whole matching section. Thereby, a global similarity value *scoreGlobal* is defined for each cut edge pair at hand respectively (see Fig. 8.10).

The global scoring was motivated by characteristics, which are often shown by typewritten documents. For example, the sections marked black in Fig. 8.10 signalize situations, where a line of text or a text block exists without continuation on the other shredder strip. This case is "punished heavily" by global scoring since this

Fig. 8.10 Step A2 of the matching situation *aligned*. You see here the global scoring of the matching situation aligned, whereby (**a**) is a suitable and (**b**) a not suitable strip pair

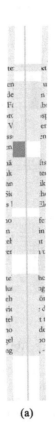

(a)   (b)

situation is an indication of two strips not suitable. However, the shown sections marked dark grey in Fig. 8.10 signalize situations, where a text block is found on one strip, which shows only partial continuation on the other strip. If a text block is not continued completely (if e.g. only two text lines of three text lines of a text block are continued), these not-continued sections are only "punished lightly", because it is not known if these sections—if the cut edge matching at hand is correct—have developed due to word wrapping or—if the cut edge matching at hand is not correct—would normally have to be continued.

On the basis of the similarity scores *scoreLocal* and *scoreGlobal*, the similarity score *scoreOverall* is determined, which specifies the level of correspondence of local and global characteristics of the cut edge pair at hand.

8.4.2.3 1:1-Matching Situation *Not-aligned*

In the case of *not-aligned*, the *1:1-matching* is divided into the following two steps: (*NA1*) determination of "meaningful" vertical orientations of the cut edges to be matched, and (*NA2*) *1:1-matching* for the orientation determined in (*NA1*), including the selection of the "best match" in terms of *1:1-matching*.

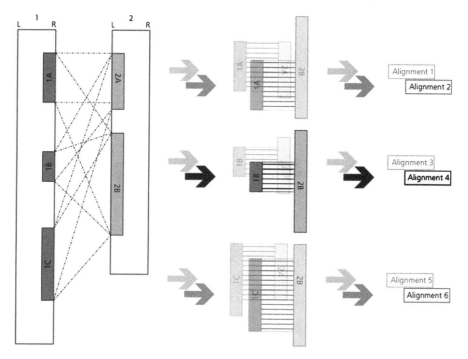

Fig. 8.11 Step *NA1* of the matching situation *not-aligned*. Determination of a vertical orientation of two cut edges (*1R, 2L*) towards each other by means of pairwise matching of LPOIs of *snippets 1* (*1A, 1B*, and *1C*) and *snippets 2* (*2A* and *2B*)

During step (*NA1*), meaningful orientations of the cut edges to be compared are calculated on the basis of the respective object and colour feature strings by means of dynamic programming algorithms (see Fig. 8.11).

In step NA2, the cut edges for each orientation determined in (NA1) are fixed vertically accordingly. Hereby, depending on the individual orientation, an according matching section develops, which is involved by the cut edge match commonly. In accordance with the proceedings of the *1:1-matching situation aligned*, for each matching section a similarity score *scoreOverall* is calculated (see Fig. 8.12). The calculated similarity score maximum specifies the level of correspondence as well as the orientation correlated to this value, i.e. the "best" orientation, of the cut edge combination at hand.

8.4.3 Shredder Assistance System

During the past years, accompanying the development of automated systems for the reconstruction of Stasi documents torn by hand, a high-capacity stock of programming modules for the assistance-based reconstruction of shredded documents has

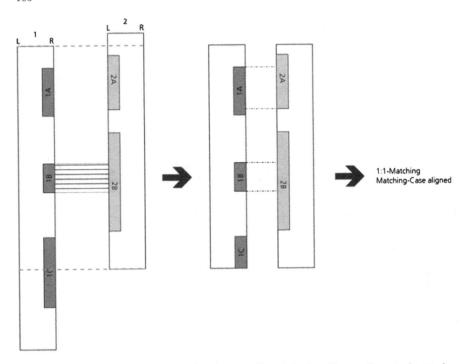

Fig. 8.12 Step NA2 of the matching situation *not-aligned*. During this step the cut edges to be compared are fixed vertically towards each other in each case according to the orientation calculated in step *NA1*. Hereby, a common matching section develops, which is evaluated according to the 1:1-matching proceedings for the situation *aligned*

been implemented at the Fraunhofer IPK. The aim of the developed software package is the virtual reconstruction of shredded paper strips to complete pages with the support of a human operator. For these means, the proceedings of phases 2 and 3 described in Sect. 8.4 are executed fully automatic by the software. The hereby determined results are—according to phase 4 of Sect. 8.4—presented to the human operator for revision in a GUI (*Graphical User Interface*), which is also used for the actual interaction with the system. Another GUI serves the control of all steps of phases 2 and 3.

For the efficient processing of large amounts of data of real application scenarios, the assistance system has been implemented into client/server architecture. In general, the system has been constructed in such a way, that different users working on different networking workstations, so-called reconstruction clients, can work on one or more reconstruction-projects at the same time. The projects as well as the respective data—i.e. the digitized strips, temporary interim result images, realized steps of the reconstruction process, etc.—are stored centrally in a SQL database. Furthermore, all feasible modules as well as the respective project-specific parameterization are on the server. If the server is made up of multiple computers or of a computer with multiple processors, a parallel processing of shredder strips is carried out, whereby performance is increased immensely.

Fig. 8.13 GUI *Preprocessing*

Additionally, there is the possibility of inserting new images of digitized shredder strips into the reconstruction client. This reflects the fact that reconstructed material often is not digitized in the beginning or mistakes in digitalization have to be eliminated afterwards.

Below, the two reconstruction client GUIs, which control the assistance system and carry out the actual reconstruction, are described.

8.4.3.1 GUI *Preprocessing*

By means of the GUI *Preprocessing*, all processing steps of phases 2 and 3 are coordinated. Ranging from the segmentation of shredder strips out of digitalization to their 1:1-matching, all individual steps of phases 2 and 3 can be parameterized by this GUI and the implementation of selected single steps can be initiated, if required. In general, the whole process chain is "kicked off" by this GUI and the system runs through it fully automatic.

For the control of individual selected steps, the following operating elements are at hand: *Project Information, Additional Information, Foreground Extraction, Noise Reduction, Feature Calculation, Clustering* and *Matching* (see Fig. 8.13).

8.4.3.2 GUI *Interactive Viewer (IAV)*

Through the GUI *IAV* different functionalities for the display of digitized shredder strips and strip pairs are made available. Also, the assistance-based virtual reconstruction of shredded documents is carried out.

Prerequisite for the virtual reconstruction by means of the GUI *IAV* is a processing of the whole process chain by the GUI *Pre-processing* beforehand. After these preprocessing steps, all pairwise matching scores for the shredder strips to be reconstructed are available in the database. Now, a user may connect to this database

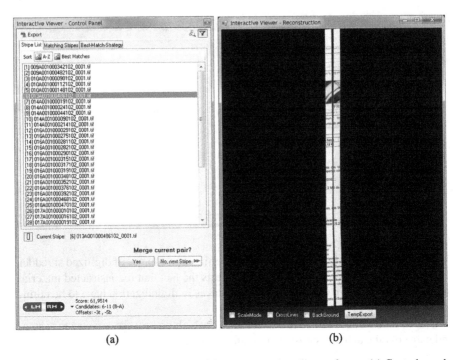

(a)   (b)

Fig. 8.14 GUI *IAV*: program interfaces of the reconstruction client software. (**a**) Control panel, (**b**) display of strip pairs

from any reconstruction client by means of the GUI *IAV* and carry out the actual virtual reconstruction of shredder strips interactively.

Figure 8.14 shows two program interfaces of the reconstruction client GUI *IAV*. The window *Control Panel* on the left serves the control of the reconstruction process and the window *Reconstruction* on the right serves the visualization of reconstruction proposals of the system on the basis of the pairwise matching scores calculated beforehand.

The interactive reconstruction is performed as follows: If the strip pair proposed by the assistance system, which is visualized in the window *Reconstruction*, is correct, the user will acknowledge this with one click on the respective button (Fig. 8.14a: "Merge current pairs: yes"). If the strip pair proposed is not correct, this is also acknowledged accordingly (Fig. 8.14a: "Merge current pairs: No, next stripe"). Thereupon, the system presents the next probable matching candidate in the window *Reconstruction*.

In the case of a positive acknowledgement, the two displayed strip images as well as the respective back sides are linked within the data base, i.e. thereafter the strips are virtually merged and will only be shown to the user as one shredder strip, forming a partial reconstruction. However, on the image level these two strips are not merged in order to enable the system to carry out *Undo*, if e.g. later on in the process it turns out that two strips in the middle of a supposedly fully reconstructed page have been reconstructed faulty. In such a case, all existing partial reconstructions can be "torn apart" again virtually by simply modifying the respective link within the database.

After a virtual merge of strip pairs has been completed, the next strip pair proposed by the system includes three or more strips: the formerly positively acknowledged partial reconstruction as well as one further strip or another partial reconstruction, which has been positively approved during another step carried out. The user checks again if the reconstruction candidate is correct or not correct. In this way, the individual strips are put together gradually to growing partial reconstructions up to complete documents (see Fig. 8.15)—if the "right candidates" exist within the system at all, which is not always the case in real application scenarios.

Figure 8.16 illustrates selected functionalities of *IAV*. One additional function has to be highlighted, which enables the user to have a look at the best candidates out of the full quantity of all strip and partial reconstruction combinations calculated by the assistance system. The candidates are sorted into descending order according to their matching score of the individual pair (see Fig. 8.16, top right). The user simply has to choose the tab page *Best-Match-Strategy* on the interface of the window *Reconstruction* and decide if the individual combination proposed on the basis of the respective list of candidates and visualized within the window *Reconstruction* is correct or not.

Compared to the course of action described above, which is based on the step-by-step processing of individual strips or partial reconstructions respectively, this course of action has one great advantage: At first, the system solely proposes promising combinations with according high matching scores. We are talking of "promising" combinations, because the probability of a pair with a high matching score being correct is of course higher than the one for a rather low matching score. Besides that, each matching acknowledged positively reduces the amount of remaining combinations on the list of candidates. Hereby, it can be a reduction of one or a few or even many candidates. This depends on the specific characteristics of the strips to be reconstructed since all possible combinations calculated, which involve one of the reconstructed border areas, are deleted from the list (see Fig. 8.15).

Hence, large datasets may be reconstructed very efficiently by means of the *Best-Match-Strategy*. This is of great help especially at the beginning of a reconstruction task, which often involves thousands of strips, and leads to a significant increase of performance.

8.5 Conclusion and Outlook

The assistance system for the virtual reconstruction of shredded documents developed at Fraunhofer IPK is a powerful tool, enabling the user to reconstruct information, which has been believed to be lost for ever since they often could not be reconstructed by hand. Now this information may be used again.

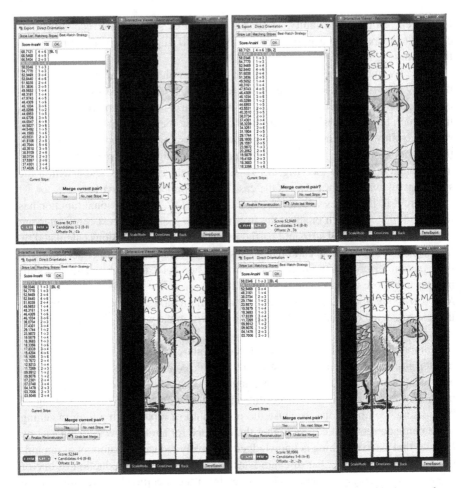

Fig. 8.15 Successive reconstruction with *IAV*. The four images show the individual steps of reconstruction within *IAV*. Departing from a partial reconstruction made up of on strip pair (*top left*) up to a partial reconstruction consisting of five strips (*bottom right*). Furthermore, the reduction of potential reconstruction candidates is illustrated. The amount of remaining strip combinations within the lists of candidates (shown on the *left* of *each image*) is reduced immensely with every merge

In the past, numerous enquiries from within the framework of tax and murder investigations could be solved successfully by means of the described technology, which enabled the evaluation of procedural relevant documents, which had been destroyed purposely. Hereby, in several projects ten thousands of paper stripes have been reconstructed respectively. Without virtual reconstruction, the information stored on this potential evidence would have been lost.

At the moment, the manual effort for the separation and digitalization of strips is still very high, making the reconstruction of large amounts of data still very time-consuming. Hence, the emphasis on further development of the technology, which

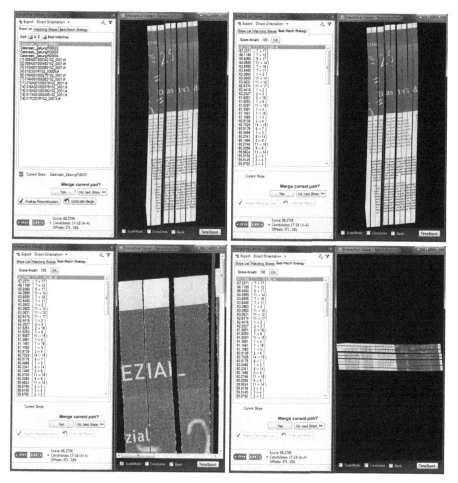

Fig. 8.16 Selected functionalities of *IAV*. The reconstruction may be carried out by either starting with one particular strip (*top left*) or from the "best" reconstruction candidates (*top right*). Furthermore, for the analysis of details, one can zoom into a proposed matching (*bottom left*) and the orientation of shown candidates may be adjusted in 90° steps, if required (*bottom right*)

forms the basis of the assistance system, has to be put on more efficient separation and digitalization solutions in order to make a time- and cost-efficient processing possible. Also, it is being worked on mathematical solutions, which are supposed to speed up the gradual interactive reconstruction of strips by determining not only single strip combinations, but chains of strips belonging together.

Chapter 9
In-Memory Technology Enables History-Based Access Control for RFID-Aided Supply Chains

Matthieu-P. Schapranow and Hasso Plattner

Abstract Modern RFID implementations leverage competitive business advantages in processing, tracking, and tracing of fast-moving consumer goods. Current implementations suffer from security threats and privacy issues, because RFID technology was not designed for secured data exchange. In emerging global RFID-aided supply chains the need for open interfaces between business partners can be abused to derive business secrets.

We developed an access control mechanisms based on in-memory technology to protect business secrets in real-time. In contrast to traditional access control mechanisms that support only bivalent access rights, our history-based access control derives concrete access rights by analyzing the complete history as well as enforcing latest possible access rights. In-memory technology is the key-enabler to handle the steady increasing query history while keeping response time latency low.

In a two-month period, more than 34 million tablets were seized, including fake antibiotics, anti-cancer, anti-malaria and anti-cholesterol medicines, painkillers and erectile dysfunction medication *(Intellectual Property Crime Report 2008/2009, 2009)*.

9.1 Introduction

The quote above reflects the tremendous number of pharmaceutical counterfeits detected at borders of the European Union. With rising efforts toward globalization in the pharmaceutical industry, mutual trust between supply chain participants becomes increasingly fragile. The automatic exchange of product and tracing data in EPCglobal networks improves business processes, but carries the risk of business

M.-P. Schapranow (✉) · H. Plattner
Enterprise Platform and Integration Concepts Chair, Hasso Plattner Institute,
August-Bebel-Str. 88, 14482 Potsdam, Germany
e-mail: schapranow@hpi.uni-potsdam.de

H. Plattner
e-mail: office-plattner@hpi.uni-potsdam.de

J. Krüger et al. (eds.), *The Secure Information Society*, DOI 10.1007/978-1-4471-4763-3_9,   187
© Springer-Verlag London 2013

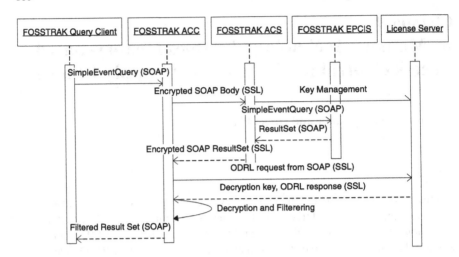

Fig. 9.1 UML sequence diagram: integration of history-based access control in FOSSTRAK

secrets exposure. Due to limited knowledge of potential business partner and the lack of access control, companies have a valid concern about the protection of their business secrets.

EPCglobal networks have competitive advantages for daily business processes by enabling automatic exchange of product-relevant data. For example, a dedicated service provider that validates the product's path automatically can perform anti-counterfeiting during good's reception.

In this chapter, we introduce an access control mechanisms that combines the following aspects:

- Real-time evaluation of the query history to specifically adapt access rights,
- Extension of bivalent access control techniques to enable a continued interval of access control, e.g. partial access,
- Evaluation of indirect rules to prevent exposure of data artifacts to derive business secrets, and
- Latest possible enforcement of access rights.

Actors of our developed History-Based Access Control (HBAC) prototype are depicted in the UML sequence diagram in Fig. 9.1.

The rest of the chapter is structured as follows. Presenting a case study of the pharmaceutical supply chain draws our motivation. We introduce technical components of EPCglobal networks, discuss selected related security threats, and present common approaches in current access control mechanisms. Selected aspects of in-memory technology as key-enabler are described and research results of our HBAC prototype are discussed.

9.2 Case Study: Counterfeits in Pharmaceutical Supply Chains

RFID technology has been named the successor of existing tracking techniques such as scanning of one-dimensional barcodes (White et al. 2007). Using RFID tags results in a number of advantages, e.g. tags can be read without establishing a direct line of sight, multiple tags can be read simultaneously, and they can cope with dirty environments. The logistics sector is one of the first implementers to guarantee traceability of fast-moving goods, such as life-saving pharmaceuticals, blood preservations, or organ donations. Tracking goods is an important factor for participants in global supply chains, i.e. RFID technology supports keeping goods moving on the road instead of keeping them in costly stocks (Schlitter et al. 2007). Compared to existing semi-automatic solution, e.g. scanning of barcodes, the implementation of RFID technology reduces time to process incoming and outgoing goods at all involved intermediate stations by enabling automatic product identification (Bovenschulte et al. 2007).

Pharmaceutical counterfeits introduce the risk of harming human-beings, e.g. when applying wrong doses, containing invalid or missing active ingredients or poisonous combinations for people with certain risks (Bos 2009). In the context of global pandemic infections, such as pandemic influenza type H1N1 in 2009 or H5N1 in 2008, the impacts of counterfeits became visible (World Health Organization 2009). Illicit drug use has been a major problem in the U.S. for years, e.g. approx. 20 million people have used illicit drugs in 2007 and more than 20 percent between the age of 18 and 20 have contributed to this statistic (Barthwell et al. 2009).

In terms of intellectual rights and property management new aspects of product tracking such as counterfeit detection become relevant. Upcoming regulations will force producers, retailers, and pharmaceutical business partners to be responsible for products showing their company logo or involvement. Tracking of their products through the entire supply chain becomes necessary. A reliable tracking mechanism is the first step in fighting counterfeits of pharmaceutical products.

Studies show that expensive pharmaceuticals, such as anti-cancer drugs and drugs treating AIDS suffer from product counterfeits with increasing rates. But also generic products are more than ever subject to plagiarism.

Already in 2006, Pfizer reported experiences with RFID-based implementations to guarantee authenticity of its Viagra pills (U.S. Pharmaceuticals Pfizer INC 2006). These activities underline the ambition of pharmaceutical manufacturers to validate the use of RFID technology as a possible way to protect their products.

In 2004, it was estimated that more than 500 billion USD were traded in counterfeits, i.e. seven percent of the world trade in the same period (International Chamber of Commerce 2004). It is argued, that this equals an increase of 150 billion USD compared to 2001 (Staake et al. 2005). In contrast, in the same period the worldwide merchandise trade increased only by approx. 50 billion USD.

At this point, it is important to highlight that estimations about the monetary impact of counterfeits vary drastically. This fact underlines that only a small number of counterfeits can be detected nowadays and that the number of unreported cases is hard to derive. Technical improvements in counterfeit detection and goods protection help to increase the amount of detected cases by implementing new barriers to prevent counterfeits entering large markets. However, anti-counterfeiting comes with the need for automatic exchange of product meta data. We consider data protection and data security concerns as increasingly important factor for the industry that demands fast and reliable security enhancements for existing EPCglobal components.

9.2.1 European Union

The EU consists of 27 member states since its last extension in 2007 when the youngest members Bulgaria and Romania joined. The EU population covers approx. 500 million citizens, i.e. approx. 7.5 percent of the world's population. Yearly, approx. 30 billion packages of pharmaceuticals are produced for this market, whereas 15 billion pharmaceuticals are only available by prescription (Müller et al. 2009a).

In 2007, a total of 43,671 counterfeit cases with approx. 80 million involved articles were reported. In contrast, a total of 49,381 counterfeit cases (an increase of 13 percent) with approx. 180 million involved articles (an increase of 125 percent) were reported in 2008 (European Commission Taxation and Customs Union 2009). A fraction of 6.5 percent of all reported cases and approx. five percent of all articles were associated with the pharmaceutical sector. Besides the categories CDs/DVDs and cigarettes, the pharmaceutical sector holds the third place according to growth rates of intercepted articles. The European Commission reports an increase of 118 percent for pharmaceutical counterfeits detected at EU borders in 2008 compared to 2007.

This increase is related to the quotation at the beginning at the chapter. By operation *Medi-Fake* more than 30 million pharmaceutical counterfeits were detected in autumn 2008 at the borders of the EU. More than 90 percent of intercepted articles were suspicious in terms of trademark infringement. More than 50 percent of all articles were intercepted during import procedures, whereas most articles were detected in air transportation. The category of life-style drugs is reported to be number one regarding detected counterfeits.

India is named as the top source of counterfeit pharmaceutical products contributing more than 50 percent of all detected articles (Shukla and Sangal 2009). This development has been constant for years. The example of India shows that counterfeiters in countries with lower law regulation benefit from pandemic diseases, such as influenza H1N1 in 2009, because consumers tend to buy pharmaceuticals preventively via the Internet (World Health Organization 2009).

9.2.2 United States

The United States Federal Food and Drug Administration (FDA) documented more than 21 counterfeit cases between 2001 and 2003 (Food and Drug Administration

2004); in 2004 this number almost tripled with 58 confirmed cases (Food and Drug Administration 2005). In contrast to this development, in 1997 to 2000 the number of detected counterfeits did not exceed six per year. This outlines two aspects. On the one hand, the number of pharmaceutical counterfeits is on the increase. On the other hand, counterfeit detection methods are being continuously improved, which results in the detection of former undetected counterfeits.

9.3 Components of RFID-Aided Supply Chains

In traditional communication networks, physical cables establish the connection between peers. If these wired links are shielded and secured against physical access, attackers are normally unable to gain unrecognized access. Although wired communication emits a low level of electronic field, which can be used to reconstruct the transported data, radio communication as used in EPCglobal networks is more exposing. Wireless communication in general is hard to secure, because data is transmitted through the ether, which can be accessed unrecognized by eavesdroppers. Current research activities address the radio interface to prevent unauthorized data. We refer to these activities as device-level security since their primary purpose is to control access to the device by unauthorized third parties. Once event data has been captured, it is stored in decentralized event repositories of individual supply chain parties. Event repositories come with another challenge: how to control the access to event data and at any level of detail? We refer to security mechanisms addressing this purpose as business-level security.

9.3.1 RFID Tags

RFID tags consist of the following components: antenna, integrated circuits, data storage, and optional equipment, such as sensors. These small radio devices can be distinguished accordingly to (a) the operating frequency band, (b) the type of tag, and (c) its read-write capabilities.

The available radio band for RFID communication is defined by standardization and influenced by country-specific restrictions (International Organization for Standardization 2004–2010), e.g. 13.56 MHz in the High Frequency (HF) or about 900 MHz in the Ultra HF (UHF) band. As comparison, current FM radio broadcasting operates in the frequency band 87.5–108 MHz. Nowadays, UHF tags with their operating frequency comparable to cellular phones are mainly in use in EPCglobal networks.

The type of tag describes the tag's capabilities. Keeping production costs low is a major requirement of passive RFID tags in EPCglobal networks for near field communication (Jones and Chung 2007). Tags can be classified accordingly to their design as passive, semi-passive, or active. Passive tags are not equipped with an autonomous power supply, i.e. passive tags need an external stimulus for operation. In contrast, active tags can operate w/o external stimulus due to their integrated power supply. As a result, their range of functionality is broader, e.g. logging of equipped sensors data such as temperature or humidity. Semi-passive tags do not include a power supply, but they are equipped with sensors that can be read when in range of a RFID reader device.

Furthermore, read/write capabilities of RFID tags can be used to classify tags additionally. Three classes of tags exist according to their read/write capabilities: (a) read-only, (b) write-once, read-many, and (c) write-many, read-many (Jones and Chung 2007). Read-only tags are a subset of write-once read-many tags, but the tag's manufacturer initializes its content. In contrast to write-once, the first user, typically the good's manufacturer initializes read-many tags. Write-many, read-many tags are equipped with a small flash storage that can be written various times.

9.3.2 RFID Reader

RFID reader devices consist of the following components: a set of antennas and a controller device. The latter implements radio interface protocols to communicate with RFID tags in a standardized way. Antennas are used to send radio signals to tags and to receive data.

9.3.3 Object Name Service

The Object Naming Service (ONS) is a yellow page service for RFID-aided supply chains (EPCglobal Inc. 2008a). It defines for a given Electronic Product Code (EPC) a mapping to the event repository of the first handling supply chain participant, typically the product's manufacturer. The inquirer can contact the EPCIS of the manufacturer to obtain further details about the product and subsequent participants that handled the goods.

9.3.4 Electronic Product Code Information Service

An EPC Information Service (EPCIS) is a standardized interface in between of internal event repositories and external inquirers (EPCglobal Inc. 2007). In other words, it is responsible for exchanging relevant internal data with other participants of the supply chain, e.g. to support automatic anti-counterfeiting services. As a result, the EPCIS is also involved in controlling access to event data to ensure privacy of internal data. Thus, we analyze the applicability of HBAC for EPCIS repositories later in this chapter.

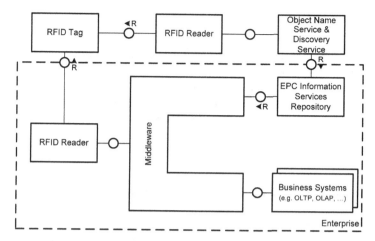

Fig. 9.2 FMC actor diagram: participant interfaces of a RFID-aided supply chain

9.3.5 *Electronic Product Code Discovery Service*

The EPC Discovery Service (EPCDS) acts as an intermediate for querying parties that pre-processes data from various EPCIS repositories and performs preliminary operations on them, e.g. aggregate internal events of companies (EPCglobal Inc. 2011). Since standards for EPCDS are under development by GS1 no concrete definitions are as yet available.

9.3.6 *Middleware*

The RFID middleware is an abstract component fulfilling a set of common tasks within a company to integrate event data in existing business systems (Müller et al. 2009b):

- Exchanging captured event data with legacy systems, such as Enterprise Resource Planning (ERP) systems,
- Unify data format of vendor-specific RFID reader devices, and
- Connect standardized EPCIS repositories.

Figure 9.2 shows an FMC actor diagram (Wendt 1991) highlighting the interfaces exposed by participants in RFID-aided supply chains. Rectangular actors such as readers and tags are connected via communication channels illustrated by circles. Enterprise infrastructure components are surrounded by a dashed line to highlight its exposed interfaces. All external components can access company's repositories via these specific interfaces. For instance, the ONS can perform queries against the EPCISs to retrieve a limited set of data. The annotated 'R' at channels indicates an one-way flow of data on this channel.

Each communication channel in the modeled architecture is a potential backdoor for attackers. External attacks can occur either when new tags are moved into the company or when data from the EPCIS is queried from external parties. The latter is a security aspect that has to be solved on the application level, e.g. by using user-specific access control lists on business level, whereas the former can only be ensured by the architecture's design decisions. Therefore, the following considerations concentrate on the weaknesses of communication between reader and tag with regards to the pharmaceutical case study.

9.4 Security Threats in RFID-Aided Supply Chains

In the following, selected security threats for RFID-aided supply chains are elaborated. These threats form the basis for competitors or attackers to obtain or manipulate product-specific data.

9.4.1 Data Protection

Protecting data against unintended access is often referred to as data protection. It involves the following aspects.

A defined set of meta data describing a certain product must remain valid through the entire product lifecycle. Although the product passes various participants in the supply chain, it must be ensured that product meta data reaches each participant in the supply chain without any manipulations or modifications. This kind of data protection is called data integrity. It ensures that no third party changes product-related data during its lifecycle.

Another aspect of data protection is data quality. It highlights the problem that data of a certain product may not be processed correctly at all times. Electronic interferences can cause that data on tag is not captured properly by the reader and contains errors or data is not read at all. All radio technologies suffer from aspects influencing the transmission quality, such as surrounding metals, long read distance, or various readers on the same limited band. Multiple readers and tags communicating simultaneously can limit coverage of radio waves and influence the reading quality.

9.4.2 Data Encryption

Communication using unreliable networks can be secured by using data encryption. *Encryption* is referred to be secure as long as no brute-force attack is able to obtain encrypted data in meaningful time. In WiFi networks encryption standards such as

TKIP or AES serve as protection for secured data exchange. Encryption is based on generating random numbers out of a large domain for encryption keys. Even the use of encryption standards does not ensure complete privacy (Beck and Tews 2008). Existing encryption standards become weak with increasing computational power. For instance, the 1977 defined Data Encryption Standard (DES) is based on a 56-bit key (Mehuron 1999). DES was replaced by its successor 3DES, because brute-force attacks with modern highly parallel Central Processing Unit (CPU) boards exposed encrypted data (Stallings 2005). In 2008, the Copacabana Rivyera project reduced the time to break DES to less than one day (Kumar et al. 2006).

For the development of radio innovations, the use of encryption sounds to be an easy way of leveraging security against the presented security threats. Although encrypting data on tags prevents attackers from reading the EPC in clear text, it does not solve issues such as cloning, spoofing of entire tags, and replay attacks. The encrypted tag content covers the same characteristics as its unencrypted EPCs, i.e. it still acts as unique identifier to identify a specific tag. In the pharmaceutical industry, a tag holding encrypted data can be tracked as reliable as a tag holding unencrypted EPC data. In contrast to communication networks, the content of the encrypted data stream is not of interest to attackers. Ultimately, encrypting the tag's content neither prevents reading nor cloning of tags and is therefore ineffective in terms of securing RFID communication.

9.4.3 Corrupted Tags

Manipulating products or placing counterfeits into RFID-aided supply chains does not necessarily involve direct attacks against reader devices. Often it is easier to manipulate the behavior of tags, e.g. by sending manipulated data. Security threats of corrupted tags are discussed in the subsequent sections.

A metallic cave can be used to shield tags from magnetic fields emitted by readers. With the help of a special wireless interface a recorded tag transmission can be transmitted to an incoming reader stimulus. In the context of the pharmaceutical case study manipulated tags can be used to exchange an authentic product by a counterfeit. Once the data footprint of an authentic pallet was recorded, it can be replayed multiple times. In the worst case, rather than being integrated into the pallet itself, it is sufficient that a manipulated tag is sent with a pallet of authentic products, e.g. it can be placed somewhere in the transportation vehicle (Juels et al. 2003). This escorting tag can query all authentic tags during goods transportation and store the results on it. When the corrupted tag is removed the content of all original tags can be read out and reused during replay attacks.

9.4.4 Corrupted Reader Devices

Data protection issues in RFID-aided supply chains are comparable to their pendants in wireless communication networks. Commodity readers can be used to read nearby moving tags. Attackers are able to access tags integrated in passports, driver's licenses, or on products packages (Koscher et al. 2009). Additional security mechanisms need to be introduced to strengthen RFID technology against read attempts by manipulated reader devices.

In the context of the pharmaceutical case study corrupted reader devices involve the following issues. EPCs can be obtained without having direct physical access to the pharmaceutical or the equipped tags. The EPC can be used to map goods and customers to create movement profiles. An advanced attack can also be used to identify the product, e.g. by using the EPC manager information with an ONS to lookup the producer of a certain good. These examples show the risk of privacy issues, e.g. enabling tracking of goods and its owned persons equally.

A possible attack scenario is to read the EPC of a blister packet after the buyer has left the pharmacy. The attacker is using a commodity reader device at the exit of the pharmacy. Each customer leaving the pharmacy is scanned for the pharmaceuticals purchased. Tags, which have not been disabled at the Point of Sale (POS), will be detected because the EPC is nowadays stored unencrypted. Tags cannot signal read attempts; the only possible way to ensure its privacy from the consumer's point of view is to destroy or shield the tag permanently before leaving the pharmacy. This ensures that no eavesdropper can read the tag content afterwards.

However, destroying tags limits process improvements after leaving the POS. For instance, a destroyed tag cannot be used to support the processing of product recalls or product exchanges. Nowadays, product recalls in the pharmaceutical sector are handled by endowing a set of pharmaceuticals with a batch number to identify affected products in this batch. With the help of unique EPCs, it is possible to identify products more accurately in case of recalls. These recalls involve expensive business steps, but RFID technology is expected to reduce these costs. Therefore, a reliable way of shielding tags against read attempts or playing tags in a hibernate mode are considered to be an effective way to ensure privacy.

9.4.5 Cloning and Spoofing

Today, low-cost passive tags are neither equipped with computational power nor security enhancements. Cloning is the process of creating a complete physical copy of an existing tag including the contained EPC. Copying an entire tag is the basis for creating counterfeits and placing them into the supply chain. Once an existing product is replaced by a counterfeit a cloned tag inherits the virtual product history of its original tag pendant.

Spoofing in terms of RFID—also known as masquerading or identity theft—exploits a trusted relationship between two peers (Hossein 2006). It is the process of manipulating a reader to assume that the EPC of a specific tag is read while the original tag is absent. Either copying the original tag or creating virtual non-existing EPCs can achieve this.

The EPCglobal standard defines that the EPC identifies a product uniquely, but it does not define how to generate its serial number component (EPCglobal Inc. 2010).

If serial numbers are assigned in a strictly linear order attackers can guess available serial numbers. To reduce the possibility of serial number guessing a proper random number generator has to be used for creation of serial numbers. This way, spoofed EPCs can be detected in a reliable way, e.g. if details for an invalid EPC are retrieved from the vendor's EPCIS, which contains a list of all valid EPCs only. In this case the response indicates that special product treatment is necessary.

9.4.6 Man-In-The-Middle Attacks

The Man-In-The-Middle (MITM) attack has been a known variant in communication networks for years (James et al. 2008). Data between two peers A and B is transmitted through different routes. Routes are dynamically setup by the underlying communication protocol. Once an attacker is controlling one node on the route between both peers or is able to influence the connection setup through a specific route, the network traffic between A and B can be captured (Hwang et al. 2008). MITM attacks are often used to gain login information for replay attacks. In the pharmaceutical case study, an attacker can use a MITM attack to acquire the tag's EPC at the POS of the pharmacy, e.g. by placing a corrupted reader device near the checkout.

At the POS in the pharmacy the kill password is sent to disable the tag of a pharmaceutical product. The business process requires that the pharmacist authenticates the pharmacy to the supplier of the product. It checks whether the given pharmaceutical is authentic and sends a query for the tag-specific kill password. The kill password ensures the consumer's privacy by preventing a match of products and consumers after having left the pharmacy.

In communication networks Intrusion Detection Systems (IDSs) are an established way to detect and monitor malicious activities (Werlinger et al. 2008).

The detection of read attempts becomes very important, especially when using security-enhanced tags. The EPCglobal standard defines a kill password that can be issued by sending a freely programmable 2×16 bit kill password (EPCglobal Inc. 2008b). Programming hundreds, thousands or hundreds of thousands of tags with a dedicated kill password is a time-consuming operation.

Executing the kill password disables the tag irreversibly, but it might be important for an attacker to obtain it. Generating unique kill passwords and maintaining the goods mapping is a challenging task. Every tag has to be programmed with a unique kill password that has to be stored in a company meta data repository mapping it to the corresponding EPC. It has to be ensured that a strong random number generator is used for kill password generation, i.e. duplicate kill passwords have to be omitted.

In the context of the pharmaceutical example an attacker could use such an obtained kill password to disable tags of original medicines and compromise their readings while faked products in the same batch are equipped with cloned tags. This way counterfeits can be transported along with authentic products without being recognized by reader gates, e.g. by duplicate reads of cloned tags or missing products in a batch when using spoofed tags.

9.4.7 Replay Attacks

In terms of communication networks replay attacks are prepared by MITM attacks (Menezes et al. 1996). Once login details have been gathered, the attacker re-uses them at later login attempts, e.g. to validate a reader against the tag in mutual authentication protocols (Schapranow et al. 2010b). Once an attacker has gathered correct authentication credentials, they can be re-used for later authentication with various tags, if identical credentials have been used. Thus, the protected tag content can be compromised after exposing authentication details of a single tag. A reliable way to protect communication systems against replay attacks is the use of One-Time Passwords (OTPs) (Russell and Gangemi 1991). The OTP is updated every time it was used, e.g. after a successful login step.

An indicator for replay attacks in RFID-aided supply chains is the presence of a certain tag passing the same read gates multiple times, e.g. hours or days after the first time passed. A manifestation in the pharmaceutical case study is a package of pharmaceuticals, which is detected at the same location of a retailer multiple times at different days.

9.4.8 Signal Interferences

Signal interferences are issues, which prevent proper tag reading, e.g. a jammer, which emits a signal with higher amplitude that blankets the tag's signal. As any kind of radio communication the quality of RFID communication depends on the used frequency band. If multiple communication attempts occur simultaneously transmission quality degrades. Various liquids and metals, such as lead or aluminum, tend to shield radio waves. Therefore, RFID technology uses company-specific radio bands for communication depending on, purpose and equipment.

Besides omnipresent uncontrolled interferences which are generated by electronic devices with improper shielding, it is possible to blanket RFID communication by triggering controlled signal interferences with higher transmission power than the original signal. For example, attackers can hide the existence of counterfeited drugs by provoking controlled signal interferences.

9.5 Access Control Mechanisms for Business Data

We understand the term access control as all efforts to limit various actions A to sensitive resources R to a certain user U. Thus, access control can be defined as triplet $(a, r, u) \, \forall a \in A, \, r \in R, \, u \in U$ as depicted in Fig. 9.3. In context of EPCglobal networks, we focus on EPC event data as resources that needs to be protected, i.e. $R = \{EPC \ events\}$.

Discretionary Access Control (DAC) describes a class of mechanisms that control access by leaving the access decision to the user. In other words, once a certain

Fig. 9.3 Entities in access
control

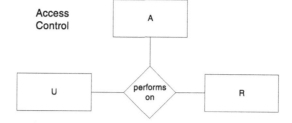

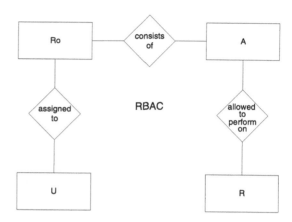

Fig. 9.4 Entities in
role-based access control

user is granted access to a resource, she/he is able to add also further users to access
the resource. Even if there is only a limited access defined for a certain resource,
e.g. read-only, the user is able to create a copy of the resource and grant individual
access to further users. For example, the operating system Microsoft Windows incorporates Access Control Lists (ACLs) while Unix uses owner-group-other-flags
for controlling access to files are representatives of DAC.

The counterpart of DAC is referred to as Non-Discretionary Access Control
(NDAC) and so the access is not directly controlled by the user, but by a dedicated
administrative entity only.

Role-Based Access Control (RBAC) controls access to resources by controlling
actions A performed by users U on resources R. In contrast to traditional access
control, RBAC groups allowed actions A in roles Ro as depicted in Fig. 9.4. As
a result, there is no direct mapping between resources R and users U. Formally, a
single role and its access rights in RBAC can be understood as $(A_{ro}, R_{ro}) \times U_{ro}$, $ro \in$
Ro with $A_{ro} = (\{a \in A | \text{permitted by } ro\}$, $R_{ro} = \{r \in R | ro \text{ is granted access to}\})$ and
$U_{ro} = (\{u \in U | \text{assigned to } ro\})$.

In contrast, Rule-Based Access Control (RuBAC) defines a set of rules Ru consisting of predicates P that are evaluated specifically when a user u is accessing
a certain resource r. Formally, it can be represented as set of $T = \{P(a, r, u, v)\}$,
$\{v \in V | \text{additional decision data}\}$ as depicted in Fig. 9.5.

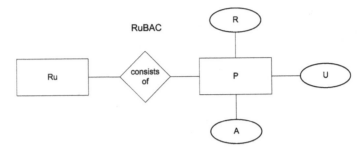

Fig. 9.5 Entities in rule-based access control

Fig. 9.6 Enterprise
performance in-memory
circle (EPIC)

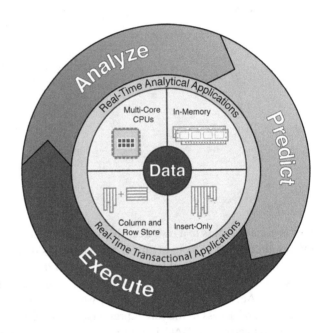

9.6 Foundations of In-Memory Technology

In the following section, we introduce selected technology building blocks used
for our HABC prototype. We refer to in-memory technology as a toolbox combining various technology aspects to enable processing of enterprise data in real-
time storing them in main memory only (Plattner and Zeier 2011). This includes
the processing of hundreds of thousands of individually queries in sub-second response time. In-memory technology enables decision making in an interactive way
without keeping redundant or pre-aggregated data. The interaction between data, in-
memory building blocks, real-time applications, and business processes is depicted
in Fig. 9.6. The outer circle describes the business perspective whereas the inner
circle depicts the technology perspective. In-memory technology helps to combine
both perspectives.

9.6.1 Combined Column and Row Store

To support analytical and transactional workloads, two different types of database systems have evolved. On the one hand, database systems for transactional work-loads store and process every day's business data in rows, i.e. attributes are stored side-by-side. On the other hand, analytical database systems aim to analyze selected attributes of huge datasets in a very short time (Schapranow et al. 2010a).

If complete data of a single row needs to be accessed, storing data in a row format is advantageous. For example, when comparing details of two customer queries, all database attributes of these two queries, such as inquirer's name, time, and content need to be loaded. In contrast, columnar database layout benefit from their storage format, when a subset of attributes needs to be processed for all or a huge number of database entries. For example, summing the total amount of products that have passed a certain reader gate involves the attributes date and business location while ignoring the EPC and the business step. Using a row store for this purpose would result in processing of all attributes of the event list, although only two attributes are required. Therefore, incorporating a columnar store is advantageous since only relevant data needs to be accessed.

9.6.2 Insert-Only

Insert-only also known as append-only describes how data is managed when inserting new data. Traditional database systems support four data operations, i.e. inserting new data, selecting data, delete data, and updating data. The latter two are considered as destructive since original data is no longer available after its execution. In other words, it is neither possible to detect nor to reconstruct all values for a certain attribute; only the latest value is available. Insert-only enables storing the complete history of value changes and the latest value for a certain attribute. This is for instance also the foundation of all bookkeeping systems to guarantee transparency. For HBAC, insert-only forms the basis to store the entire history of queries for access decision. In addition, insert-only enables tracing of access decisions, which can be used to perform incident analysis.

9.6.3 Lightweight Compression

Compression in the context of in-memory technology refers to a storage representation consuming less space. A columnar storage supports the use of lightweight compression. Due to the fixed data type per column, subsequent values are within a given interval, e.g. integer values. In addition, the given data type defines an upper threshold of individual values. Depending on the source of data, the concrete number of individual values is lower, i.e. the amount of distinct values. A small representation of data requires only the amount of distinct values. For example, all incoming queries are stored in a logging database table for processing by the HBAC. If ten supply chain participants query details for the same product, it traditionally results in ten-times the same query. Instead of storing the query redundantly dictionary compression could be employed. For example, the query is stored once and mapped to a small integer representation. Within the database table only the integer value is stored and queries are translated to use the integer representation. The real-world value for the integer representation is replaced just before returning the result set to the client. As a result, the database performs all operations on compressed data without decompressing the data. This compares favorably to the uncompressed format, which requires transferring ten times the complete query through the memory hierarchy of the server. Transferring only the compressed data increases cache usage while at the same time it lowers cache misses.

9.6.4 Partitioning

In the following, we distinguish two partitioning approaches: vertical and horizontal partitioning. A combination of both approaches is also possible.

Vertical partitioning refers to rearranging individual database columns. It is achieved by splitting columns of a database table in two or more column sets. Each of the column sets can be distributed on individual database servers. This can also be used to build up database columns with different ordering to achieve better search performance while guaranteeing high-availability of data. Key to success of vertical partitioning is a thorough understanding of the application's data access patterns. Attributes that are accessed in the same query should be located within the same partition since the operations locating and joining of data may degrade overall performance.

In contrast, horizontal partitioning addresses long database tables and how to divide them into smaller chunks of data. As a result, each piece of the database table contains a subset of the complete data within the table. Splitting data in equivalent long horizontal partitions is used to support search operations and improve scalability. For example, a scan of the request history results in a full table scan. Without any partitioning a single thread needs to access all individual history entries to check the selection predicate. When using a naïve round robin horizontal partitioning across 10 partitions, the total table scan can be performed in parallel by 10 simultaneously processing threads reducing response time by approx. 1/9 compared to the single threaded full table scan. This example shows that the resulting partitions depend on the incorporated partitioning strategy. For example, rather than using round-robin as partitioning strategy attribute ranges can be used, e.g. inquirers are portioned in groups of 1,000 with the help of their user id or the requested EPC.

9.6.5 Multi-core and Parallelization

Parallelization can be achieved at a number of levels in the application stack of enterprise systems—from within the application running on an application server to query execution in the database system. As an example of application-level parallelism, let us assume the following: Incoming queries need to be processed by EPCIS repositories in parallel to meet response time thresholds. Processing multiple queries can be handled by multithreaded applications, i.e. the application does not stall when dealing with more than one query. Threads are software abstractions that need to be mapped to physically available hardware resources. A CPU core can be considered as single worker on a construction site. If it is possible to map each query to a single core, the system's response time is optimal. Query processing also involves data processing, i.e. the database also needs to be queried in parallel. If the database is able to distribute the workload across multiple cores a single system works at its optimal. If the workload exceeds the physically available capacities of a single system, multiple servers or blades need to be involved for work distribution to achieve optimal processing behavior. From the database perspective, partitioning datasets supports parallelization since multiple cores across servers can be involved for data processing.

This example shows that multi-core architectures and parallelization depend on each other while partitioning is the basis to use resources in parallel.

9.6.6 Any Attribute as Index

Database tables are stored as collections of tuples. However, accessing data within these collections can be improved by certain techniques. The most common types of organizations are heaps, ordered collections, hashed collections, and tree indices. In contrast to traditional indices, such as B-trees, the group-key index stores the encoded value as a key and the position of the corresponding value as a value list. The group-key concept allows increasing the search performance for transactional and analytical queries significantly.

9.6.7 Active and Passive Data Store

We define two categories of data stores: active and passive. We refer to active data when it is accessed frequently and updates are expected, e.g. access rules. In contrast, we refer to passive data when this data either is not used frequently and neither updated nor read. Passive data is purely used for analytical and statistical purposes or in exceptional situations where specific investigations require this data. For example, tracking events of a certain pharmaceutical product that was sold five years ago can be considered as passive data. Why is this feasible? Firstly, from the business perspective, the pharmaceutical is equipped with a best-before data of two years after its manufacturing date, i.e. even if the product would handled now, it is no longer allowed to be sold. Secondly, the product was sold to a customer four years ago, i.e. it left the supply chain and has been typically already used within this time-span. Therefore, the probability that details about this certain pharmaceutical are queried is very low. Nonetheless, law regulation requires the tracking history to be preserved, e.g. to prove the used path within the supply chain or when selling details are analyzed for building a new long-term forecast based on historical data.

This example gives an understanding about active and passive data. Furthermore, introducing the concept of passive data has the advantage to reduce the amount of data, which needs to be accessed in real-time, and to enable archiving. As a result, when data is moved to a passive data store, they no longer consume fast accessible main memory and thus free hardware resources. Dealing with passive data stores involves the need for a memory hierarchy from fast, but expensive to slow and cheap. A possible storage hierarchy is given by: memory registers, cache memory, main memory, flash storages, solid state disks, SAS hard disk drives, SATA hard disk drives, tapes, etc. As a result, rules for migrating data from one store to another need to be defined, we refer to it as aging strategy or aging rules. The data aging process, i.e. migrating data from a faster storage location to a slower one, is considered as a background task. This task occurs on a regular basis, e.g. weekly or daily. Since aging involves reorganization of the complete data set, it should be processed during times with low data access, e.g. during nights or on weekends.

9.6.8 Reduction of Application Layers

 In application development, layers refer to levels of abstractions. Each application layer encapsulates specific logic and offers certain functionality. Although abstraction helps to reduce complexity, it also introduces obstacles. The latter result from various aspects, e.g. (a) functionality is hidden within a layer and (b) each layer offers a variety of functionality while only a small subset is in-use.

From the data's perspective, layers are problematic since data is marshaled and unmarshaled for transformation in the layer-specific format. As a result, the identical data is kept in various layers redundantly. Moving application logic to the data it operates on results in a smaller application stack and therefore code reduction. Furthermore, reducing the code length also results in improved maintainability and an improved use of hardware resources.

9.6.9 History-Based Access Control

We developed a special purpose access control mechanism that involves the analysis of the complete history before making access decisions. In the following, we outline limitations of existing access control mechanisms and show why in-memory technology is the key-enabler for HBAC. We share results of our research prototype and indicate its applicability to existing EPC event repositories.

9.6.10 Existing Limitations

Traditional access control mechanisms have major limitation in context of the presented pharmaceutical case study. The traditional decision making process is characterized by a result in the set decision → {decline, grant}. It is reasonable if data that need to be secured consists of a single value, e.g. "15" or "house". In terms of the pharmaceutical case study, data that need to be secured are single events stored in the EPCIS repository. A single event e can be considered formally as quadruple consisting of $e = (epc, t, loc, ba)$, with epc describing the product's unique EPC, t describing the timestamp when e occurred, loc describing the location where e was recorded, and ba describing the associated business action while e occurred (Schapranow et al. 2011). In terms of integrity of EPCglobal networks it is necessary to share major information about a certain good with all involved parties. However, some information may be exposed selectively or hidden as defined by EPCglobal (EPCglobal Inc. 2007). We consider the event data as sensitive data that need to be secured since they can be used to derive business secrets.

Therefore, we consider an access decision as an interval instead of a set: "decision → [decline, grant]". As a result, this new definition of an access decision offers a continuous spectrum of partially granted or declined access. In other words, various fragments of an event can be controlled and accessed when considering each event as atomic unit of control.

Furthermore, traditional access control mechanisms make it hard to maintain all combinations of possible inquiring parties, when inquirers are not known beforehand. For example, maintaining all possible inquirers in a global pharmaceutical supply chain is impossible for a single supply chain participant as the high maintenance overhead would exceed potential benefits. Although RBAC reduces the administration of individual parties to roles, it does not reduce the maintenance overhead to assign data and corresponding roles.

RuBAC supports the definition of access rights for an unknown set of inquirers in a better way by defining arbitrary predicates that involve any external factors. Evaluating these predicates helps to find appropriate access criteria. However, the definition of rules involves the knowledge what data to protect. In context of the pharmaceutical case study, the confidential business data is not exposed directly, but it can be derived through semantic combination of responses for other queries.

In contrast to traditional access control mechanisms, HBAC combines multiple access control mechanisms.

Firstly, to reduce the complexity of handling individual inquirers a user group concept is incorporated. Users are grouped whereas access rights are assigned to these groups explicitly. As a result, it is possible to share different levels of detail with different user groups.

Secondly, sets of freely definable rules are maintained. Each of the rule sets is individually proofed; each rule can either be permitted or prohibited. For rule evaluation the complete query history of the current inquirer will be analyzed. In this way, data that has been exposed days, weeks, or even month ago is used in the decision making process to prevent the exposure of business secrets. All rules of a single rule set are combined via a logical AND operation and applied. Since business relationships are hard to define, these rules are defined indirectly, i.e. not for a concrete inquirer or date. They are more abstract and describe sets of information that should never be exposed together regardless of the way they are accessed. As a result, even partially exposed data can be secured, due to the fact that only data that was already granted access to first will be available.

Thirdly, combining the user group relationship and the outcome of the rule sets results in a filtering operation. For example, the XML-encoded EPCIS result is filtered and data is partially removed or replaced by default values. As a result, a valid XML-encoded EPCIS response is generated while the exposed data are controlled by our history-based access control automatically.

9.6.11 In-Memory Technology as Key for History-Based Access Control

We consider access control as a time critical action. All subsequent tasks involving controlled data build on the performance of the incorporated access control mechanism since it defines the minimal processing time of all processes. HBAC creates a special challenge for modern computer systems, e.g. due to the analysis of the steady increasing history of queries. Thus, we incorporate in-memory technology as key-enabler for HBAC. In the following, we apply selected in-memory building blocks to achieve real-time response time for HBAC.

Due to the characteristics of the query history, it can be considered as a Write-Once Read-Many (WORM) store. New entries are appended to the end of the list while the list increases steadily. The technical representation of this WORM store is an insert-only database table of the incorporated in-memory database. Storing all data in a columnar format instead of the traditional row format comes with the following benefits.

Firstly, storing columns side by side is the basis for applying lightweight compression techniques. For example, if identical queries are performed multiple times whenever in the history, it is only necessary to store the query once. All subsequent identical query entries will only reference to the initial query by using a short reference identifier. This technique is referred to as dictionary compression and excludes the need to decompress the data when working with it. Data need to be remapped only once when it is returned to the application. As a result, the storage requirements of the history table can be reduced by factors 10–100. Further, the data buses and CPU caches of the computer system are used more efficiently. Due to the lower storage demand of a single query, more queries can be transferred via the same bus without extended bus capacity. In contrast, traditional row layout stores all data of a row side by side that contain various columns and different data types. These row stores make any lightweight compression techniques hard to apply.

Secondly, the columnar database layout supports horizontal and vertical partitioning of the history. We used vertical partitioning to distribute certain partitions of the entire history to individual blades, CPUs, and cores. With the help of the partitioning rule, the time can be tuned to detect relevant history entries.

The use of highly distributed system architectures benefits from a high number of individual database partitions. With the increasing number of available CPU cores in modern computer systems, the amount of partitions can be increased. Each core performs the search of relevant entries on a subset of the entire query in parallel and returns the sub result to create the overall result. In other words, if the processing of the history exceeds the desired two-second response time threshold, additional CPU cores can be added and the number of partitions increased.

Another important aspect of partitioning is to distinguish between active, often used and passive, less often used data. Consider a partitioning rule that moves all products that are older than a one year time since manufacturing to a passive partition. The passive partition can be stored on any kind of slow archive storage since it stores data only due to law regulations. As a result, the data within the fast active store is kept minimal and data follows a lifecycle, i.e. data is cleaned-up after a certain period.

9.6.12 Keeping the Entire Query History

Assume a bookkeeping system that protocols all account activities. When a wrong booking occurs, it requires a corresponding booking correction instead of deleting the incorrect booking. As a result, the account's history represents a complete list of all executed transactions. HBAC builds on the same principle: it keeps the complete query history, i.e. query, timestamp, and querying party. As a result, it is possible to travel through the time to figure out who sent a specific query.

Let us consider the following rule: we want to ensure that two information iA and iB are never exposed together. For example, iA gives details about supplier A and iB gives details about a certain product that is shipped by your company using ingredients from supplier A. In terms of business relations your company has an interest to keep details about A from buyers of B to protect its suppliers and business relationships.

HBAC ensures the protection of business relationships by analyzing the complete query history. If an inquirer is asking for iB it is typically a person that bought B. In terms of anti-counterfeiting, any supply chain party that handles B will query for iB to ensure it is an authentic product. However, if one of these inquirers starts to query for iA, this is untypical behavior. For example, an end consumer that bought B is now querying for iA. To prevent exposure of the combined information the complete history is analyzed. It is detected that iB was queried first, i.e. partially the rule is fulfilled. In combination with the current query for iA both parts of the rule are fulfilled and the result set will be adapted. Since iB was queried first, one can assume that this information was also exposed first. As a result, any queries for iB will be satisfied at any time as this 'secret' can be considered as exposed. Due to the fact that iA was considered as occurring later than iB, the former 'secret' is still secured, when iB is exposed. Alternatively, if the exposure order is reversed, iB remains secured.

This example shows that the exposure order has a direct impact on current access decisions. In other words, the complete history needs to be traversed in chronological order when performing a concrete access decision. In-memory technology builds the foundation to process the history in acceptable time to implement a real-time access control mechanism.

9.6.13 Applying History-Based Access Control to Existing Repositories

Our developed research prototype aims to support both enhancing security of event repositories while enabling a flexible integration. From our perspective, a flexible integration of security enhancements is essential when dealing with existing IT infrastructures. During the design of the HBAC prototype we already focused on a non-disruptive integration. Thus, services of existing event repositories are not affected by applying HBAC. To verify our design, we exemplarily combined the EPCIS repository of the Free and Open Source Software for Track and Trace (FOSSTRAK) with our HBAC.

The FOSSTRAK EPCIS repository offers via HTTP a capture and via SOAP a query interface. In the following, we focus only on the query interface since it is the source for existing event data where data access needs to be controlled. We can assume that queries and result sets are exchanged as XML message encapsulated in the SOAP message body.

For enabling transparent integration we created a client server component for access control as depicted in Fig. 9.1. The Access Control Client (ACC) and its pendant the Access Control Server (ACS) are FOSSTRAK-specific. The ACC extends the existing FOSSTRAK query client by performing data encryption and decryption tasks and enforcing access rights by filtering result sets. The ACS is located on the event owner site and extends the functionality of the EPCIS repository by performing data encryption and decryption tasks for queries and result sets. As a result, the ACC/ACS pair ensures confidentiality and integrity of requests and event data. Furthermore, the ACC performs access control and result set filtering on client site. The interested reader may ask why the result set is not filtered on the server site before returning the event data to the client. The design of the security extension follows the principle of very late access control. In other words, although an attacker was able to acquire encrypted event data the access to this data can be declined remotely. Access rights are enforced with the help of specific licenses. The license contains the decryption keys for the encrypted result set and the client-specific access rights defined in the Open Digital Rights Language (ODRL). With the help of the ODRL rules the filtering of event data is performed by the ACC on client site before returning the decrypted and filtered event set to the FOSSTRAK query client. The described design decision keeps the load for access enforcement on server site low and eliminates the need for adaption of the EPCIS software.

The interested reader may argue that performing decryption on client site increases the risk of tampering. However, when considering participants of global pharmaceutical supply, all valid participants are interested in keeping integrity of the overall supply chain. As a result, the ACC is the digital identity of an inquirer and increases mutual trust. In addition, when tampering with the client software is detected the licenses for accessing event data can be revoked immediately by the event owner while the identity of the leakage is known. From a business perspective, it is feasible to require all participants of a global supply chain to have a valid certification for integer use of IT systems. In mutual business relationships this certificate can be used to identify new business partners that are allowed to make use of ACC.

9.6.14 Research Results

Adding security extensions comes not for free. To understand the concrete costs in terms of latency, we performed benchmarks to obtain results for the architecture's throughput. We focused on the processing of the query history since its length is steadily increasing. In the following, we focus on varying the factors: length of query history, number of querying parties in the history, and history partitioning.

Figure 9.7 depicts the overall response time of HBAC running on a MySQL database system with a random number of inquirers. The response time increases dramatically within the first few 1,000 queries up to approx. 23 s shortly before a history length of 4,000 entries. For a few hundred entries in the history the response time remain less or equal to two seconds, thus we stopped the benchmark with a history length of 10 k entries.

In contrast to MySQL, Fig. 9.8 depicts the response time behavior when using in-memory technology for a history length of more than 125,000 entries. Data is partitioned across 10 equally sized ranges. After a steady increase of response time for the first few 1,000 entries up to 0.45 s, the response time stabilizes around less

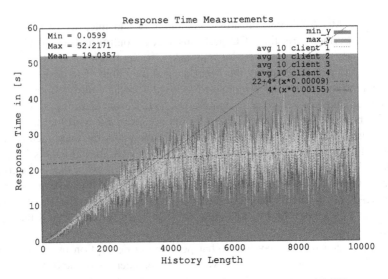

Fig. 9.7 Impact of history length on response time, random 10 k inquirers, MySQL

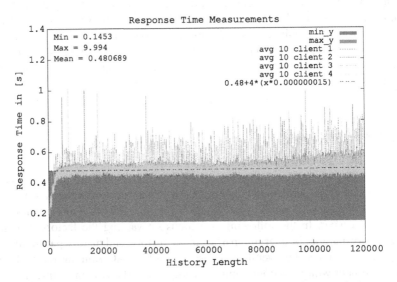

Fig. 9.8 Impact of history length on response time, random 10 k inquirers, partioning across 10 ranges, in-memory database

than 0.5 s for the remaining progress. It shows the advantage of in-memory technology while processing data. The performance advantage results from the in-memory building blocks as described above.

9.7 Summary

In our work, we developed an access control mechanism that extends traditional static control systems by real-time analysis of the query history. In-memory technology is the key-enabler to make this vision come true. Our requirements to keep a complete list of chronological queries for all supply chain parties and to perform real-time processing of it is also very hard to achieve even with modern computer systems. We introduced selected aspects of in-memory technology that build the foundation for a highly parallelized and scalable system landscape to reduce processing time. Furthermore, in-memory technology reduces the requirements for storing the steady increasing history by lightweight compression methods.

We used the developed in-memory prototype to extend the functionality of the open-source EPCIS repository FOSSTRAK. Our implementation shows that adding security features is possible while keeping the mean response time below one second per query. Furthermore, our implementation proves that security features can be integrated in existing systems in a transparent way without modifying any third party source code. We consider our contribution as a major step to automatically protect business secrets in modern supply chains. Checking the history of all queries is the best input to reconstruct what data was already exposed and to verify whether additional responses would result in exposing a business secret.

References

Barthwell, A. G., Barnes, M. C., Leopold, V. R., & Wichelecki, J. L. (2009). *National survey on drug use and health*. Arlington: Center for Lawful Access and Abuse Deterrence.

Beck, M., & Tews, E. (2008). Practical attacks against WEP and WPA. http://dl.aircrack-ng.org/breakingwepandwpa.pdf. Accessed 31 Oct 2011.

Bos, J. V. D. (2009). Globalization of the pharmaceutical supply chain: what are the risks?—The FDA's difficult task. *Society of Actur, 61*, 24–26.

Bovenschulte, M., Gabriel, P., Gaßner, K., & Seidel, U. (2007). RFID: prospectives for Germany—the state of radio frequency identification-based applications and their outlook in national and international markets.

EPCglobal Inc. (2007). EPCIS standard 1.0.1. http://www.gs1.org/gsmp/kc/epcglobal/epcis/epcis_1_0_1-standard-20070921.pdf. Accessed 31 Oct 2011.

EPCglobal Inc. (2008a). EPCglobal object name service standard 1.0.1. http://www.gs1.org/gsmp/kc/epcglobal/ons/ons_1_0_1-standard-20080529.pdf. Accessed 31 Oct 2011.

EPCglobal Inc. (2008b). EPC radio-frequency identity protocols—class-1 generation-2 UHF RFID protocol for communications at 860 MHz—960 MHz—1.2.0. http://www.gs1.org/docs/uhfc1g2/uhfc1g2_1_2_0-standard-20080511.pdf. Accessed 31 Oct 2011.

EPCglobal Inc. (2010). Tag data standards 1.5. http://www.gs1.org/gsmp/kc/epcglobal/tds/tds_1_5-standard-20100818.pdf. Accessed 31 Oct. 2011.

EPCglobal Inc. (2011). Discovery services standard. http://www.gs1.org/gsmp/kc/epcglobal/discovery. Accessed 31 Oct 2011.

European Commission Taxation and Customs Union (2009). Report on EU Customs En-forcement of IP Rights.

Food and Drug Administration (2004). Counterfeit Drug Task Force Report.

Food and Drug Administration (2005). Counterfeit Drug Task Force Report.

Hossein, B. (2006). *Handbook of information security*. New York: Wiley.

Hwang, H., Jung, G., Sohn, K., & Park, S. (2008). A study on MITM vulnerability in wireless network using 802.1X and EAP. In *Proceedings of international conference on information science and security*. Washington: IEEE Computer Society.

International Chamber of Commerce (2004). The fight against piracy and counterfeiting of intellectual property. http://www.iccwbo.org/home/intellectual_property/fight_against_piracy.pdf. Accessed 31 Oct 2011.

International Organization for Standardization (2004–2010). ISO/IEC 18000: information technology—radio frequency identification for item management.

IP Crime Group (2009). IP Crime Report 2008–2009.

James, M. S., Tittel, E., & Chapple, M. (2008). *Certified information systems security professional study guide* (4th ed.). New York: Wiley.

Jones, E. C., & Chung, C. A. (2007). *RFID in logistics: a practical introduction*. Boca Raton: CRC Press.

Juels, A., Rivest, R. L., & Szydlo, M. (2003). The blocker tag: selective blocking of RFID tags for consumer privacy. In *Proceedings of the conference on computer and communications security* (pp. 103–111). New York: ACM Press.

Koscher, K., Juels, A., Brajkovic, V., & Kohno, T. (2009). EPC RFID tag security weaknesses and defenses: passport cards, enhanced drivers licenses, and beyond. In *Conference on computer and communication security* (pp. 33–42). New York: ACM.

Kumar, S., Paar, C., Pelzl, J., Pfeiffer, G., & Schimmler, M. (2006). Breaking ciphers with COPACOBANA—a cost-optimized parallel code breaker. In *Cryptographic hardware and embedded systems* (pp. 101–118). Heidelberg: Springer.

Mehuron, W. (1999). *Data encryption standard*. Technical report, National Institute of Standards and Technology.

Menezes, A. J., Vanstone, S. A., & van Oorschot, P. C. (1996). *Handbook of applied cryptography*. Boca Raton: CRC Press.

Müller, J., Pöpke, C., Urbat, M., Zeier, A., & Plattner, H. (2009a). A simulation of the pharmaceutical supply chain to provide realistic test data. In *Proceedings of the international conference on advances in system simulation* (pp. 44–49).

Müller, J., Schapranow, M.-P., Helmich, M., Enderlein, S., & Zeier, A. (2009b). RFID middleware as a service—enabling small and medium-sized enterprises to participate in the EPC network. In *International conference on industry engineering and engineering management 2*, Beijing: IEEE Computer Society.

Plattner, H., & Zeier, A. (2011). *In-memory data management*. Berlin: Springer.

Russell, D., & Gangemi, G. T. (1991). *Computer security basics*. Sebastopol: O'Reilly & Associates Inc.

Schapranow, M.-P., Kühne, R., & Zeier, A. (2010a). Enabling real-time charging for smart grid infrastructures using in-memory databases. In *1st IEEE LCN workshop on smart grid networking infrastructure*.

Schapranow, M. P., Zeier, A., & Plattner, H. (2010b). *A dynamic mutual RFID authentication model preventing unauthorized third party access*. Melbourne: IEEE Press.

Schapranow, M.-P., Zeier, A., & Plattner, H. (2011). A formal model for enabling RFID in pharmaceutical supply chains. In *44th Hawaii international conference on system sciences*.

Schlitter, N., Kähne, F., Schilz, S. T., & Mattke, H. (2007). Potentials and problems of RFID-based cooperations in supply chains. In *Innovative logistics management: competitive advantages through new processes and services* (3rd ed.). Berlin: Erich Schmidt Verlag GmbH & Co.

Shukla, N., & Sangal, T. (2009). Generic drug industry in India: the counterfeit spin. *Intellect Property Rights, 14*, 236–240.

Staake, T., Thiesse, F., & Fleisch, E. (2005). Extending the EPC network: the potential of RFID in anti-counterfeiting. In *Symposium on applied computing*, New York: ACM.

Stallings, W. (2005). *Cryptography and network security* (4th ed.). Upper Saddle River: Prentice Hall.

U.S. Pharmaceuticals Pfizer INC (2006). Anti-counterfeit drug initiative workshop and vendor display. http://www.fda.gov/OHRMS/DOCKETS/dockets/05n0510/05N-0510-EC21-Attach-1.pdf. Accessed 31 Oct 2011.

Wendt, S. (1991). *Nichtphysikalische grundlagen der informationstechnik. Interpretierte formalismen*. Berlin: Springer.

Werlinger, R., Hawkey, K., Muldner, K., Jaferian, P., & Beznosov, K. (2008). The challenges of using an intrusion detection system: is it worth the effort? In *Proceedings of the 4th symposium on usable privacy and security* (pp. 107–118). New York: ACM.

White, G., Prabhakar, G., Abdrazak, A., & Gardiner, G. (2007). A comparison of barcoding and RFID technologies in practice. *Journal of information, information technology and organizations, 2*.

World Health Organization (2009). Warning on purchase of antivirals without a prescription, including via the Internet. http://www.who.int/medicines/publications/drugalerts/Alert_122_Antivirals.pdf. Accessed 31 Oct 2011.